The Battle of Fayetteville Arkansas

April 18, 1863

The Civil War battle in Arkansas between Arkansas Confederate Cavalry and Arkansas Union Cavalry

by
Russell Mahan

2nd Edition 2019
Published by Historical Enterprises
2780 Bella Sol Drive, Santa Clara, Utah 84765
HistoricalEnterprises@Outlook.com

The Battle of Fayetteville, Arkansas

Table of Contents

The Civil War in Northwest Arkansas

In northwest Arkansas the war of 1861-1865 was not a War Between the States, with all of the people of Arkansas united for the Confederacy against Union troops from the Northern states, but rather a true Civil War. It was neighbor against neighbor without regard to geographical considerations. While a majority of people in that part of the state favored the Confederate cause, a very substantial minority was strongly Unionist in sentiment and opposed to secession. After the war it was estimated that 2,000 men from Washington County, Arkansas, served in the Confederate Army, while perhaps up to 800 were in the Union service.[1]

The people of Arkansas were by nature politically moderate. In the first round of Southern secession following the election of Abraham Lincoln to the presidency in November of 1860, Arkansas watched with interest but took no part. In February of 1861 the people of the state voted against secession in their selection of delegates to an Arkansas secession convention, which met in March and voted down disunion. But when the guns at Fort Sumter were fired, Southern moderation was the first casualty. When President Abraham Lincoln called for volunteers to put down the rebellion and asked Arkansas for a contribution of men, it not only refused but joined the newly formed Confederate States of America.

Indeed, the start of war aroused most of the Southern population into a great wave of Confederate patriotism. Men flocked to join military regiments to defend their new nation. Patrick Cleburne of Helena, Arkansas, who went on to great fame in the pantheon of Confederate heroes, explained this patriotism well in a letter:

> *I am with the South in life or in death, in victory or in defeat. I never owned a Negro and care nothing for them, but these people have been my friends and have stood [with] me on all occasions. In addition to this, I believe the North is about to wage a brutal and unholy war on a people who have done them*

> *no wrong, in violation of the constitution and the*
> *fundamental principles of the government. They no*
> *longer acknowledge that all government derives its*
> *validity from the consent of the governed. They are*
> *about to invade our peaceful homes, destroy our*
> *property, and inaugurate a servile insurrection, murder*
> *our men and dishonor our women. We propose no*
> *invasion of the North, no attack on them, and only ask*
> *to be let alone.*[2]

The regiments quickly formed and squabbled among themselves for the honor of which one ought to be granted the title of "*First* Arkansas Cavalry" or "*First* Arkansas Infantry." Arkansas men served in every major theater of the war, including in far away Virginia from Bull Run to Appomattox. Many regiments were sent to the so-called "Western Theater" of Tennessee and neighboring states, fighting in every major battle. Still others stayed in Arkansas to fight on the home front in the "Trans-Mississippi" district.

For the first few months of the new Confederacy, Unionists in northwest Arkansas grimly went about their business and hoped for the best, keeping (mostly) silent. Without having loaded their belongings into a wagon or having taken a single step down the road, they found themselves living in a new country not of their choosing. What had once been patriotic loyalty to one's country was now treasonous disloyalty, and the transition to new circumstances was difficult or impossible for many. Confederates begrudgingly tolerated their known but quieted dissent.

Beginning with the Battle of Wilson's Creek (also known as Oak Hills) near Springfield, Missouri, on August 10, 1861, things became more ominous as young men from Arkansas started dying over this disagreement. Toleration and forbearance toward dissenters became much harder to maintain by the majority. What had been minor harassment gave way to persecution, and soon to life-threatening confrontations. Union men began fleeing into the mountains, and finally, in the first half of 1862, they began making their way to th United States Army lines in Missouri.

The war in Arkansas was never given the consideration it merited. Geography, the political importance of the capitals of Washington and

Richmond, and population distribution in 1861 made it inevitable that there would be a concentration of troops in the East. Virginia was like a great center of gravity, pulling Union and Confederate regiments toward it from all parts of the North and South. Political leaders gave decreasing concern and resources the further west one went from there. West of the Mississippi River, in the so-called "Trans-Mississippi" theater of the war, both sides received the stereotypical "step-child" treatment from their national leaders. This situation affected and hampered everything that went on in Arkansas, both Union and Confederate.

The Southern victory at Wilson's Creek secured northwest Arkansas for the Confederacy, but only for a short time. From the time of secession in May of 1861 until February of 1862, Rebel authority was not seriously challenged. Then it all changed, and Civil War in all its ugliness and horror came to the area. It stayed for more than three unhappy years.

The change came when a Union army commanded by General Alexander Asboth moved southward from Missouri into northwest Arkansas in February 1862. The outnumbered Confederate forces withdrew ahead of him, deciding that the time to oppose the invasion would come later. Anxiety was running high for everyone, military and civilian alike.

As the Rebels pulled out, it was decided that the abundant army supplies that could not be moved should not simply be left to the Federals. Confederate General Benjamin McCulloch permitted his men to take away from the supplies in Fayetteville whatever foodstuffs they could carry, and the men went to it with vigor. Local citizens began to join in the taking. Before long, however, things took a nasty turn and widened into a free-for-all pillage of the entire town of Fayetteville. Stores and private homes were ransacked. Then, after the infantry withdrew, Confederate cavalry on orders from McCulloch set fire to many of the buildings in town in an apparent scorched-earth policy of destruction. Most of the downtown business square and many other structures were destroyed.

The disaster was recorded by Pastor William Baxter of Fayetteville, the Northern-born President of Arkansas College, a Protestant minister and an ardent Unionist:

Pastor William Baxter.
Washington County
Historical Society.

The next morning dawned, and in haste the officers of the retreating [Confederate] army departed. But soon bands of Cavalry... dashed into town and began firing the buildings which had been used for military purposes; some of them contained large quantities of beef and bacon, which soon added violence to the flames. Then the large stables, once the property of the Overland Mail Company, were destroyed; the steam mill, which had been furnishing the rebel army ten thousand pounds of flour per day, was consumed; and, as we gazed on the clouds of smoke pierced here and there by tongues of flame, we felt that the fate of our beautiful mountain city was sealed.[3]

Baxter went on to describe the torching of the Fayetteville Female Institute (also known as Van Horne's Academy after its founder) which contained numerous defective artillery shells from its previous use as an arsenal. First the building went up in flames, then the shells exploded and set other structures ablaze as well. In another month, Arkansas College was burned, too.

The devastation was also described by Confederate Lt. George Taylor of the 17th Arkansas Infantry:

Fayetteville presented indeed a sad spectacle when we passed through on the 20th. I could not but contrast the beautiful quiet little town of last May when we were so heartily welcomed to the then devastation and waste and ruin manifest all around. Heaven help a country where an army must linger, be it friend or foe. What Citizens now left in Fayetteville, seemed perfectly panic-stricken - seemed to be utterly regardless of anything like protection of property. Stores all along Main Street were thrown open to the Missouri and Arkansas soldiers.

> *Amid the destruction it could not but be amusing to see great heavy-bearded fellows carrying around fancy little toys - rattlers, made to amuse very small juveniles. Bonnet frames - old French flowers - nearly every man had a looking-glass.*[4]

War with all its vengeance had arrived. On February 22, 1862, Fayetteville was a town between armies. "Night came," Pastor Baxter wrote, "and southward the camp-fires of the [Confederate] armies of Price and M'Culloch could be seen, while to the northern sky a glow like that of the Aurora Borealis was given by those of the Federal soldiery."

The next day Yankee soldiers marched into Fayetteville amid the smoldering ruins and raised the Stars and Stripes atop the Washington County Courthouse. Local Union ladies saluted the flag as it went by. The home of a Northern-born, Unionist lawyer named Jonas M. Tebbetts was selected as the Union Army headquarters. Tebbetts was called "Judge" but his service on the bench had been before his arrival in Fayetteville. Tebbetts built the house in the 1850s just across Dickson Street from Arkansas College where his friend Pastor Baxter was president.

General Asboth issued a proclamation on February 28th:

> *To the citizens of Fayetteville:*
>   *Sent in command of the United States army of the Southwestern District of Gen. Samuel R. Curtis, commanding, I have occupied your town to arrest the wanton destruction of public and private property already inaugurated by the Confederate troops; to sustain those of its inhabitants who are faithful to the laws; to encourage all who may have temporarily wavered in their duty under the threats of bad designing men, and to establish the law and order essential to the public weal.*
>   *While, therefore, calling upon the loyal citizens of this town to aid me in the furtherance and accomplishment of these objects, I at the same time offer to all who may have faltered in their fealty, but who now truthfully declare their allegiance to the laws*

of the Union, the protection of its flag. Deserted firesides cannot be guarded, but every house containing a living soul shall have the protection of our power. None, therefore, should depart. Those absent should return....[5]

Arkansans who stepped forward and publicly embraced the Union cause soon regretted it. Asboth had done them a great disservice. Within days after the proclamation, he and his army withdrew from Fayetteville and marched back into southwest Missouri. Pastor Baxter immediately recognized that this exposed the Union people living in the area to danger from the Confederates. General Ben McCulloch sent Southern troops back into town, and had some of the Northern sympathizers arrested. Such is the nature of civil war. Baxter hid in the woods to avoid being imprisoned.

Judge Tebbetts was not so lucky. He was quickly arrested and taken to the General, then sent to prison in Fort Smith. Before McCulloch could do anything further, however, he was killed at the Battle of Pea Ridge. Tebbetts was released and hastened to sell the house in Fayetteville and leave the Confederacy forever. Thomas B. Van Horne, the founder of the Fayetteville Female Institute, and Reverend Robert Graham, founder of Arkansas College, native Northerners and Unionists, also fled.

The Battle of Pea Ridge (also known as Elkhorn Tavern) Arkansas, occurred about thirty miles north of Fayetteville on March 7th and 8th, 1862. Many of the casualties were brought into Fayetteville, including the bodies of Generals Ben McCulloch and James McIntosh. Although it was an important Northern victory which effectively secured Missouri for the Union, Arkansas ground was quickly abandoned as Federal lines were pulled back once again into southern Missouri.

By default, Fayetteville and northwest Arkansas were back under Confederate control, not by actual occupation by soldiers but by the default of Yankee absence. For most of the remainder of the year 1862 Fayetteville was not really occupied by either side, but rather left as something of a no-man's-land. Both sides raided, pillaged and recruited, but neither was in actual military occupation. Nevertheless, under such conditions, it was primarily Confederate territory.

On October 28, 1862, Union troops again marched into Fayetteville. This time even the local Unionists suffered. Pastor Baxter complained:

> *My yard was soon stripped of poultry, my house was filled with soldiers, and we were feeding them as rapidly as we could; some few of them, who had been down on a scouting party, knew me, and their presence was valuable for a time; but other hungry crowds came, and many of them I have no doubt thought that they were cleaning out a secessionist, and did it with a good will; and while I was doing my best to feed as many as the kitchen would hold, and sending them away to have their places filled by others, if possible hungrier still, a number went to the back part of the smokehouse, pulled off some of the planks, and appropriated everything they could lay their hands on....*
>
> *They were in an enemy's country, their supply-wagons were not always in reach, their necessities were pressing, and there was neither time nor inclination for discussing the rights of property; they had learned, too, that every man who had any thing to lose, almost invariably became a Union man on their approach....*[6]

Within two or three more days, the Federals again retreated northward. Many Union families and escaped slaves fled with them. Then on the evening of December 6, 1862, General Francis J. Herron's Federal Army marched through again, going south. Herron had sent cavalrymen in advance to post a guard at every house in town for the protection of local persons and property. The townspeople were grateful.

The following day, December 7th, a few miles southwest of town the bloody Battle of Prairie Grove took place. The sounds of the distant artillery could be plainly heard in Fayetteville that Sunday. Smoke could be seen, and even musketry heard at times. It was surely hard to think of anything else that day. What would it mean for the townspeople? The remaining Union people especially were in a condition of distress, for a Federal defeat would mean that they would likely have to flee their homes. "When the sounds of battle ceased our anxiety was not relieved," wrote Pastor Baxter. "We knew the conflict was over, but not yet whether the brave little band, which had marched by ere the sun of that day had risen, were victors or vanquished."

The initial word was that the Federals had been surprised and defeated. The next morning "my heart sank within me as I saw proof...of disaster and retreat," said Baxter. "But my gloom was soon turned to gladness. I hurried down and found that the artillery was for the defense of the town, which was to be henceforth a military post." Confederate troops, although initially successful in the battle, withdrew from the field during the night and retreated southward, evacuating the northwest part of the State. These artillery mentioned was later withdrawn prior to the Battle of Fayetteville four months later.

Then came the broken bodies of men as Fayetteville was flooded with more than a thousand battle casualties. Every place that could give shelter became crowded. Baxter wrote that "the town was one vast hospital" and described the scene at Sophia Sawyer's Fayetteville Female Seminary (which was a different institution from the Female Institute):

> [T]he entire floor was so thickly covered with mangled and bleeding men that it was difficult to thread my way among them; some were mortally wounded, the life fast escaping through a ghastly hole in the breast; the limbs of others were shattered and useless, the faces of others so disfigured as to seem scarcely human; the bloody bandages, hair clotted, and garments stained with blood, and all these with but little covering, and no other couch than the straw, with which the floor was strewed, made up a scene more pitiable and horrible than I had ever conceived possible before. Nor was this the only place which presented so sad a spectacle; it was repeated in about twenty other buildings, including the various churches, all of which were thronged with the sad wrecks of humanity from that field which in song and story will long be remembered as Prairie Grove....
> I must say, however, that the wounded bore their sufferings like the heroes that they were, no

> *abandonment to grief or useless complaining, but on the contrary, many were calm, and some even cheerful.[7]*

Dr. Seymour D. Carpenter, a 36-year-old Union officer from Cedar Rapids, Iowa, assigned to the area, reported on some of the many problems he found when he arrived:

> *There was a dearth of blankets, under-clothing, and every other article necessary for the sick. We found all the Churches, the Seminary, and the private houses crowded, the wounded lying on straw, hardly any cooking utensils, and no table service. A horrible stench pervaded all the place, and the death toll was terrible.[8]*

Another result of the battle was that a Federal post was established at Fayetteville, with Colonel M. LaRue Harrison of the First Arkansas Union Cavalry in command. The political situation of the townspeople instantly reversed itself. Local Unionists returned to their homes; secessionists fled. Some people simply changed sides - at least outwardly. Except for an outpost along the line of the Indian Territory (Oklahoma), all other Union regiments were pulled back to Missouri, leaving Fayetteville as an exposed and isolated outpost, tenuously in contact with the main army in Cassville by a telegraph wire and the Cassville Road. This was the situation that existed on April 18, 1863, when the Battle of Fayetteville occurred.

Pastor Baxter described the result of the relentless war upon the land and people of northwest Arkansas:

> *Sad, sad, however, was the change in our once beautiful and prosperous inland city; the fences had nearly all disappeared, shrubbery and fruit-trees were ruined, houses were deserted, nearly all of the domestic animals killed, dead cavalry-horses lay here and there; the farms, for miles around, were laid waste, the fences having been used to keep up the hundreds of campfires which were seldom permitted to go out by night or day; stables were pulled down, outbuildings burnt, and the very spirit of destruction seemed to rule the hour. The contrast drawn between now and better days was most*

painful; all that was once valued was destroyed or defaced, and, worse than all, the future seemed to have no promise....

I had pursued my labors as minister and teacher long after others had abandoned them, and I felt my work there was done. True there were a few faithful ones of the flock to which I ministered, and yet I felt that necessity must soon compel even these to leave, as the means of living were becoming more limited daily; few farmers would attempt to raise a crop in such an unsettled state of affairs, the mills were nearly all destroyed, and the food question was fast becoming one that engrossed the attention of all - one, too, which none cold solve satisfactorily.[9]

Confederate General William L. Cabell also described northwest Arkansas in the spring of 1863. "There is nothing north of the [Boston] mountains to subsist either men or horses; nothing growing and nothing remaining of their last year's crop...."

This was the Civil War in northwest Arkansas as it moved toward the coming Battle of Fayetteville.

The Arkansas Confederates

The most remarkable thing about the Battle of Fayetteville was that it occurred on Arkansas soil between Arkansas Confederates and Arkansas Unionists. It was therefore a small but representative sample of the Civil War in general.

The Confederate commander in the battle was Brigadier General William Lewis Cabell. He was born in Virginia on the first day of 1827. He attended West Point and graduated in 1850, ranking 33rd in a class of 46. While stationed at Fort Gibson in the Indian (Oklahoma) Territory in 1856, he married Harriet A. Rector of Fort Smith, Arkansas.

Brigadier General
William L. Cabell
Library of Congress.

As soon as war began in April of 1861, Cabell resigned his commission in the United States Army and tendered his services to the Confederacy. He was appointed as a Major and assigned as quartermaster for the forces gathering in northern Virginia under General Pierre Beauregard. It is said that Cabell helped Beauregard design the famous Confederate battle flag. As quartermaster, he called upon the women of the South to donate their red and blue silk dresses to be made into flags.

In January of 1862 Cabell was sent to Arkansas where he was again assigned to be quartermaster. In April he went to Mississippi and was placed in command of a brigade. He was badly injured on October 5th when his horse stumbled and fell, landing on his left leg. He returned to Arkansas to recuperate.

On October 26, 1862, Major Cabell wrote to the Adjutant General of the Confederate States requesting promotion to the rank of lieutenant colonel. For some reason, now seeming to be unjustified, the Confederacy was very slow in promoting Cabell. In his letter he pointed out that he entered the Confederate Army before either his native state of Virginia or his adoptive state of Arkansas had seceded, and that he had "feelings of wounded pride" at seeing others junior to him being promoted over him. "I am not serving for either pay or for the love of war," he said. "I am fighting as a matter of duty and nothing will or shall prevent me fro meeting the enemies of my country as long as I am able until they are driven back" and peace obtained.

Recovering from his leg injury, he returned to active service in November 1862, and was once again assigned to quartermaster duties. Then in early 1863 Cabell was at last given the long-overdue promotion to brigadier general. For the nucleus of his brigade he was assigned the two regiments of Monroe and Carroll, which he supplemented with independent companies of "partisan rangers" and whatever forces he could recruit within the state.

On March 28, 1863, three weeks before the Battle of Fayetteville, Cabell issued a proclamation calling the people of northwest Arkansas to arms, urging them to greater sacrifices and effort.

Monroe's First Arkansas Confederate Cavalry

James C. Monroe was born in South Carolina but now was from Clark County, Arkansas. He was originally the Captain of Company B of Colonel James F. Fagan's First Arkansas Confederate Infantry. He was soon elected as Lieutenant Colonel. The regiment was sent to Virginia, and on the way across the new Confederate nation attracted attention because one of its captains was Robert H. Crockett, a grandson of the famous Davy Crockett, and another was Donelson McGregor, a grand-nephew of the wife of Andrew Jackson who had been raised near Jackson's home, the Hermitage. The unit saw minor action at the Battle of Bull Run (Manassas) in July 1861, and then was transferred to Tennessee. At this point Fagan and Monroe left the regiment to return to Arkansas.

In the fall of 1862 they organized a regiment of cavalry with Fagan once again as Colonel and Monroe as Lieutenant Colonel. When Fagan

The Confederate Order of Battle

Brigadier General William L. Cabell
(approximately 900 men)

Monroe's 1st Arkansas Cavalry
- Colonel James C. Monroe

Carroll's 1st Arkansas Cavalry
- Lieutenant Colonel Lee L. Thomson

Hughey's Arkansas Mountain Battery
- Lieutenant William M. Hughey

Hill's Battalion of Arkansas Cavalry
- Captain Oliver Basham

Arkansas Confederate Partisan Rangers
- Captain J. R. Palmer's Company
- Captain William Brown's Company

Dorsey's Missouri Squadron
- Colonel Caleb Dorsey

Noble's Texas Battalion
- Lt. Col. Sebron Miles Noble
Company A, 12th Texas Cavalry
- Captain Joseph A. Weir's
Company H, 20th Texas Cavalry
- Captain Wilson J. Coggins

was soon promoted to be brigadier general, Monroe assumed the colonelcy of the regiment, which became known as "Monroe's First Arkansas Cavalry." On November 28, 1862, he was in a skirmish at

Cane Hill, not far from Fayetteville. "Here," reported the overly optimistic Colonel Thomas Ewing of the 11th Kansas Cavalry, "were killed Colonel Monroe, commanding Fagan's Texas Cavalry, Captain Martin, of the Arkansas Cavalry, and others." Monroe was not dead, and he did not command Texas Cavalry. Yielding to superior numbers, in a fighting Rebel withdrawal that day, Monroe's Arkansas regiment lost 3 men slightly wounded and 4 horses killed in that confrontation.

In the Battle of Prairie Grove, Colonel Monroe commanded a two-regiment brigade of cavalry consisting of his own regiment and that of Colonel Charles A. Carroll. On the evening of December 6, 1862, the night before the battle, Monroe made an initial charge on a Union position at the foot of a mountain. "I then ordered up my whole command," he reported afterward, "and formed them in line of battle behind the hill, out of range of the enemy's guns. I then sent two companies around to attack the enemy's right flank, and at the same time made a vigorous charge in front. The enemy held his position until we were in 10 paces of him, when he broke and fled in confusion. I would have pursued him were it not for the nature of the ground, which was so rugged that it was impossible to ride over." He lost three men killed and twelve wounded.

General Thomas C. Hindman, the commander of the Confederate Army then in northwest Arkansas, reported on the charge: "There, from the crest of the mountain to its base, about sunset a sharp engagement occurred, in which Col. J.C. Monroe and his brigade of Arkansas cavalry...greatly distinguished themselves, charging a superior force of the enemy's cavalry with boldness and vigor, breaking his ranks, and only ceasing to pursue when recalled." Brigadier General John S. Marmaduke, in his battle report, stated: "The conduct of Colonel Monroe, who charged at the head of this brigade, and of the officers and men under his command in this affair, was gallant in the extreme."

The strategy of Confederate General Hindman at the Battle of Prairie Grove was bold, indeed. Facing one enemy under General James G. Blunt, he knew another Union army under General Francis J. Herron was fast approaching from the north. Knowing he could not defeat them if they united, he decided to hold Blunt in place, strike and destroy Herron before they could join together, then turn back and defeat Blunt. It was the sort of a plan that is brilliant if it succeeds (like Robert E. Lee at Chancellorsville), or foolish if it fails (like

Thomas Hindman at Prairie Grove).

Monroe was given the critical assignment of demonstrating in front of Blunt to keep his Union Army in place, while Hindman attacked Herron. Through the night of the December 6th-7th Blunt had his men sleep in line of battle, expecting an attack from Hindman before daylight. The Yankees could see Confederate campfires in the distance, which Monroe's men kept going through the night while Hindman's army marched to strike Herron. At dawn, Monroe moved froward in a skirmish line, but that feint was all that was going to happen. Later, about 10:30 in the morning, artillery could be heard at Prairie Grove, and Blunt finally figured out that he was being held by a token force and marched away to help Herron.

"Monroe," wrote Marmaduke, "in obedience to orders, attacked the enemy at daylight on Sunday morning, and, by his daring and skill, kept the enemy in the belief, until 10:00 a.m., that the attack was to be made in that direction. Upon the enemy retreating, he pursued and formed a junction with the main force about sunset on the battle-field." Monroe reported that, "I was ordered to remain in my present position and engage the enemy at daylight. I dismounted all the men in both regiments who had long-range guns, deployed them as skirmishers, moved forward and commenced skirmishing with the enemy just after daylight, which was kept up until cannonading was heard in the direction of Fayetteville, when the enemy retired, and I moved slowly and cautiously in the direction of Cane Hill."

When Brigadier General William L. Cabell was later assigned to put together a brigade, Monroe and his men were assigned to it. The life of this regiment was succinctly stated in the April 1863 returns for Company B. "This company," it recorded, " has been constantly on the march; no opportunity for drilling until within the last two weeks. Company is improving rapidly. Armed with shot guns. No accoutrements. Clothing very poor."[10]

On the expedition to attack Fayetteville the regiment was personally commanded by Colonel Monroe. Though very much reduced from original numbers, the regiment was present in full force except for Company H, which was detached for service in Van Buren. The individual companies had been completely reorganized by elections just a few days before. The record for 2nd Lieutenant S. B. Hardin of

Company E contains the uncharitable notation, "Thrown out of office in the reorganization of the Co. on the 14th April/63." Many men were serving in new positions and under new officers as they headed into the battle at Fayetteville.

Carroll's First Arkansas Confederate Cavalry

Carroll's First Arkansas Confederate Cavalry, like Monroe's, vied for the title of "first." Carroll was an outstanding recruiter of men for the army. "In raising troops in Arkansas," General Thomas C. Hindman wrote, "Col. Charles A. Carroll was more successful than any other officer, and is entitled to high credit. He was valuably assisted by [other named officers] and put in the service three full regiments of infantry and one of cavalry."

Although the regiment continued to bear Carroll's name, it was not long before Carroll himself moved on to other things. Command then went to Lieutenant Colonel Lee L. Thomson, who led the regiment into action in the fighting at Cane Hill, Arkansas, on November 28, 1862. There a brigade consisting of Carroll's cavalry (under Thomson) and Monroe's cavalry operated under the command of Monroe. At the time that Monroe's men made the charge described above, Thomson and his regiment were able to pour a heavy fire into the Union defenders from a dismounted position almost directly to the Federal rear.

On that date, the "Carroll's brigade" (under Monroe and Thomson) muster rolls showed that there were 1,700 men. In reality, however, there were only 200 effective men and 317 that were "non-effective. The non-effective men were composed of the sick and men whose horses were in bad condition.... [T]he horses...are worn down, having been constantly on the move for six weeks, and for want of forage and shoeing." Nonetheless, despite low numbers, General Hindman reported that "Colonel Monroe and his brigade of Arkansas cavalry greatly distinguished themselves."

Thomson and his men were with Monroe on the morning of December 7, 1862, when they made a feint against Union General Blunt's forces to hold them in place while the bulk of the Confederate army attacked General Herron at Prairie Grove. That evening they re-joined the main army after the battle had concluded, and the following two days covered the retreat of the Rebel army as it moved southward from the battlefield.

Lieutenant Colonel Lee L. Thomson still commanded "Carroll's Cavalry" at the time of the Battle of Fayetteville.

Hughey's Arkansas Artillery

In December of 1861 a battery of artillery was organized in Lafayette County, Arkansas. That county is located in the extreme southwestern corner of the state and then contained the town of Texarkana. The Captain of the battery was D.W. Harris, and the First Lieutenant was William M. Hughey. Hughey was Georgia-born, 26 years old , and single. Before the war he lived in LaGrange Township of Lafayette County with two physicians.

On November 28, 1862, Hughey was with Colonel Carroll at the fight at Cane Hill, Arkansas. "Of the mountain howitzer battery attached to my brigade and commanded by First Lieutenant Hughey," Carroll reported, "only one section was serviceable." In a fighting withdrawal, the battery lost 3 men wounded and 4 horses killed. It is likely that the battery was with Colonel Monroe on December 6th-7th at the Battle of Prairie Grove.

When General Cabell formed his new brigade, Hughey went into it as a part of Carroll's regiment. Cabell soon recommended Hughey for promotion to Captain. "He is now in command of a three gun battery in my command," the General wrote on April 2nd, "and I find him competent, faithful and industrious."[11] General Theophilus Holmes approved the promotion, but the final approval by General E. Kirby Smith was not given until May 7th, after the Battle of Fayetteville. This led to some inconsistent designating of his rank in the reports as both Lieutenant and Captain.

Only two of Hughey's three guns, however, were available for the attack on Fayetteville. These were iron (as opposed to brass) 6-pounder guns that had been rehabilitated after having been previously discarded. Some of the gunners were from a recent instruction camp at Dardanelle, and apparently going into battle for the first time as artillerymen. It was typical for men from the regiment to volunteer or be detailed for detached service with the artillery, and this is likely what had happened here.

Hill's Battalion of Arkansas Confederate Cavalry

In forming his brigade, Cabell organized a new battalion under Colonel John F. Hill of Johnson County, Arkansas. At the time of the battle it consisted of five companies of men. Hill had been a Captain and then a Colonel in the 16th Arkansas Infantry, but resigned in May of 1862. By the spring of 1863 he was putting together this battalion of cavalry.

It was a tough assignment. "Hill's regiment," Cabell said later when it had grown beyond a battalion, was "raised from deserters and jayhawkers who had been lying out in the mountains and forced into service."[12] They were difficult men and unreliable as soldiers.

When the Confederates moved out for Fayetteville, three of those companies were left at Ozark because they were poorly armed and their horses were unshod. Most likely, Hill himself stayed in Ozark in command of his men there. The other two companies went on the expedition.

Irregular Arkansas Confederate Guerrillas

Other irregular units were also present with Cabell at the Battle of Fayetteville. These men were called rangers, home guards or guerrillas by the Confederates, and bushwhackers by the Federals. They were not within the formal Confederate organization, often operated independently upon their own agenda, and were alleged by Unionists to be nothing more than marauding outlaws.

The regular Confederate military did not have the manpower to properly occupy northwest Arkansas, so it was decided to "authorize" independent groups of men to function in military roles in order to maintain at least some semblance of a presence in the area. This was done under the following order:

> *General Orders No. 17*
> *Head Quarters, Trans-Mississippi District*
> *Little Rock, Ark., June 17, 1862*
>
> *I. For more effectual annoyance of the enemy*

upon our rivers and in our mountains and woods, all citizens of this district, who are not subject to conscription, are called upon to organize themselves into independent companies of mounted men or infantry, as they prefer, arming and equipping themselves, and to serve in that part of the district to which they belong.

II. When as many as ten men come together for this purpose, they may organize by electing a captain, one sergeant and one corporal, and will at once commence operations against the enemy, without waiting for special instructions. Their duty will be to cut off Federal pickets, scouts, foraging parties, and trains, and to kill pilots and others on gun-boats and transports, attacking them day and night, and using the greatest vigor in their movements. As soon as the company attains the strength required by law, it will proceed to elect the other officer to which it is entitled. All such organizations will be reported to these head-quarters as soon as practicable. They will receive pay and allowances for subsistence and forage, for the time actually in the field, as established by affidavits of their captains.

III. These companies will be governed in all respects by the same regulations as other troops.

Captains will be held responsible for the good conduct and efficiency of their men, and will report to these head-quarters from time to time.

By command of Major-General Hindman.[13]

The Federals did the very same thing, calling for Unionist men to operate in irregular units to roam the countryside looking for Rebels. These authorizations constituted a license for armed bands to roam the countryside, for good or evil, under the sanction of the law.

It appears that at least two Confederate independent companies, those of J. R. Palmer of Washington County and William "Buck" Brown of Benton County, were with Cabell in his attack on Fayetteville. Federal Colonel Harrison reported that those of Peter Mankins and James Ingraham were also there, but this is not reflected in Confederate documents.

Missouri and Texas Cavalry

The people of Missouri were terribly divided by the Civil War - even more so than in Arkansas. The state did not secede, though it was represented by a star on the Confederate flag. The Union victory at the battle of Pea Ridge permanently secured Missouori for the Union, but savage guerrilla war and raids from Arkansas kept the state in constant turmoil.

Many thousands of Missourians formed into Confederate military units, one of which was commanded by 28-year-old Caleb Dorsey of Pike County, Missouri. He was born into a prominent Maryland family, but his father came west in the 1830s. Dorsey was involved early in the war as a member of the First Missouri State Guard under General Sterling Price. Later, he commanded a losing, outnumbered fight at Mt. Zion Church in Boone County on December 28, 1861.

On February 15, 1862, Dorsey was captured by Union forces near the Osage River, and held as a prisoner of war. He was released from prison in Fort Warren, Massachusetts, on July 31st for the purpose of being exchanged for a Union officer of equal rank held by the Confederates.

Returning to Missouri, Dorsey was quickly back in the saddle in command of "Dorsey's Squadron." This unit varied considerably in size, but at the time of the Battle of Fayetteville it consisted of only a company or two.

Dorsey's Missourians had already had a minor entanglement with the First Arkansas Union Cavalry. On Sunday, January 25th, the Federal Cavalry had captured a steamboat called the *Julia Roane* on the Arkansas River with 175 sick and wounded Confederates on board. Since the ship was used only as a hospital, the patients were paroled and it was permitted to proceed. Confederates derided this capture of the infirm as a cowardly attack.

To prevent it happening again, the Confederates organized a riverside escort. On February 2, 1863, Colonel Dorsey and about thirty cavalrymen were assigned as the escort. At White Oak, seven miles west of Ozark, Federal Captain Galloway attacked. Dorsey counter-charged and drove them back, with only a few casualties on each side.

Because of its remoteness on the southwestern frontier, Texas faced no threat of invasion from the Union. Having more soldiers that were needed for its own protection, Texas could and did send many of its regiments eastward to help elsewhere. A number of them ended up in Arkansas.

Lt. Col. Sebron Miles "Sebe" Noble commanded the Texas cavalrymen who were present at the Battle of Fayetteville. He was Mississippi-born, single, 29 years old, and had been a land agent in Nacogdoches before the war. In August of 1861 he enlisted as a private in Company H of the 4th Texas Cavalry, and worked his way up the chain of command. In March of 1862 he became Captain of the newly organized Company A of the 17th Texas Cavalry, and later was promoted to Lieutenant Colonel of that regiment.

The Texas men present at Fayetteville under Noble's command consisted of Captain Joseph P. Weir's Company A of the 12th Texas (Parson's) Cavalry, and Captain Wilson J. Coggins' Company H of the 18th Texas (Darnell's) Cavalry.[14]

The Arkansas Federals

On the afternoon of Monday, May 6, 1861, an intense political excitement was sweeping Arkansas. The moderation which had prevailed in the councils of government and in the hearts of most of the people was blown away in cannon fire at far away Fort Sumter and replaced by the virtual unanimity which only war can produce. They very men who had voted to remain in the Union just a few weeks before at the first session of the Arkansas Secession Convention were now on the verge of unopposed secession.

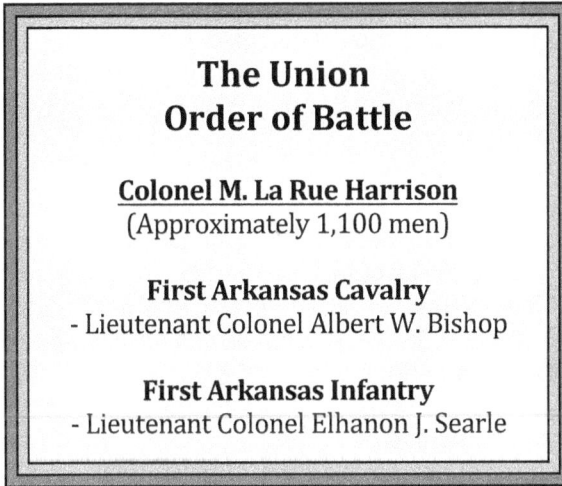

The Union
Order of Battle

Colonel M. La Rue Harrison
(Approximately 1,100 men)

First Arkansas Cavalry
- Lieutenant Colonel Albert W. Bishop

First Arkansas Infantry
- Lieutenant Colonel Elhanon J. Searle

Those same convention delegates were gathered for a second session at the chambers of the State House of Representatives in Little Rock. It its original voting in early March, the convention had voted down three secession resolutions by votes of approximately 40 to 35. The people of Arkansas themselves (at least those voting) had rejected secession on February 18th by a vote of 23,626 to 17,927. Now, the convention was recalled to the task in the wake of war in South Carolina, disunion was resisted only by five delegates. David Walker of Fayetteville, president of the convention and previously a Unionist himself, called for a unanimous vote for secession.

The political pressure on the five remaining Union delegates was unbearable. Standing in the path of a tidal wave of Confederate enthusiasm was more than most men could do, especially when to do so called forth the bitterest denunciations of treason from erstwhile friends and associates. It was no wonder that four of the five opponents changed their votes in favor of the secession they could not stop.

Then all eyes turned upon the fifth dissenter, an educator from Huntsville named Isaac Murphy. He rose to his feet to speak. "I have cast my vote after mature reflection, and have duly considered the consequences," he boldly explained, "and I cannot conscientiously change it. I therefore vote 'No.'" Amid a storm of verbal abuse and scorn which instantly burst forth from the other representatives and the public in attendance, a woman in the gallery threw a bouquet of flowers to Arkansas' last Unionist delegate. It was among such people as these - a man with his vote and a woman with her flowers - that the First Arkansas Union Cavalry had its beginning.[15]

In the months that followed, many Union men in Arkansas had experiences similar to a young man by the name of Thomas Wilhite. His story is told here as a representative sample of what men like him endured. In June of 1861, about a month after secession, six men rode up to the Wilhite farmhouse at Fall Creek, about 21 miles south of Fayetteville. Wilhite, 24 years old, saw them coming and armed himself with a rifle and revolvers.

For some time Wilhite had by necessity been extremely cautious. As an unabashed Union man in a Confederate state, he had learned to keep a careful watch to his rear. Working in the fields, he kept a rifle slung over his shoulder. Sleeping at night, he not only kept a rifle at the head of his bed but sometimes wore pistols around his waist as well. Men had occasionally come to arrest him at his own farm, but finding the plow without its owner, they would leave the equipment unmolested in the eerie belief that they were at that moment within the deadly aim of the unseen Wilhite.

Upon the arrival of his six visitors, Wilhite came around the corner of the house, taking them by surprise. They informed him that they had come to arrest him for being a Union man. Holding his rifle on them, Wilhite replied that they had better get reinforcements for the job as

Lieutenant Thomas Wilhite.
Prairie Grove Battlefield State Park.

six men were not enough. He cautioned them against going for their guns, as he would take at least one man with him for sure, and there was no telling just which man that would be. The visitors reconsidered the situation, then asked that they be permitted to leave unharmed. Wilhite generously granted their request.

On a different occasion, another group of men came to the farm. Finding their intended victim away from home, they consoled themselves by showing their hanging rope to Wilhite's hapless mother.

One day Thomas Wilhite went over to a mill owned by T.K. Kidd near Cane Hill on a business errand. He found himself surrounded by a men who had the objective, as usual, of arresting him. Armed, as always, Wilhite was armed, and told them he had 13 shots, and that he would not be taken alive. The crowd mellowed at this point, and disbanded when Mr. Kidd suggested that Wilhite might yet make a good Southern soldier. He eventually would make a good Southern soldier, but not in the way Kidd intended.

Attending a Baptist gathering one Sunday, Pastor Thomas Dodson launched into a torrent of abuse against Union sympathizers. "If there is a Union man within the sound of my voice," he declared, "I want him to leave the house, and leave it now-a." Wilhite, in his unrestrained and unintimidated way, headed for the door. "Then go-a," the minister intoned, "and darken not again the house of God."

By November of 1861 Wilhite had all that he could take. He felt compelled to go into hiding, taking refuge in a cave in the Boston Mountains south of Fayetteville which he had previously supplied with corn and forage for just such an eventuality. There he stayed with several horses, only occasionally returning home, and then with the greatest of caution.[16]

The Union victory at the battle of Pea Ridge, Arkansas, in March of 1862 considerably dampened Confederate aspirations in Missouri and encouraged the many Unionists in northwest Arkansas. Such men began to work their way northward to Federal Army lines. By May Thomas Wilhite was himself ready to take up arms in defense of his nation and against the majority of his home state. He set out for the Union lines in southwest Missouri and by the time he arrived, he had thirty men with him. They would be back - armed, mounted and wearing blue.

Lieutenant Colonel
Albert W. Bishop
Washington County
Historical Society.

Lieutenant Colonel Albert W. Bishop would come to serve with these Union fugitives and to know their personal stories well. In his partisan way, he wrote in late 1862:

> *The Government knows but little of the sufferings of the loyal men of the Border. It is no easy thing to adhere to the Union in a seceded State, and when insult, outrage and beggary are the consequences..... In no section of the country has the Great Rebellion created such intense personal hate, or separated more widely friends and relations, than in the South-West....*

> *The poison [of disunionism] spread, and soon infused itself into the minds of hundreds of peaceable citizens, transforming them into bands of armed and head-strong men, ready at a moment's notice to fire the house, plunder the property, and take the life of an inoffending neighbor, if suspected, even, of sympathy with the "Lincoln Government." Nobody, in fact, could be so bad as a "Fed".... Personal abuse was followed up by the shotgun....*
>
> *To remain longer at home was worse than to leave wives and children (temporarily, as they thought) and thus began the hegira of the Southwest. About this time Federal forces were again accumulating at Springfield, and thither hunted, but not disheartened, Unionists of Arkansas bent their steps.... No obstacles daunted, no dangers appalled them. Lying in the woods by day, at early nightfall they resumed their toilsome journey, carefully shunning highways and trusting to the instinct of self-preservation...for ultimate safety....[17]*

In the spring of 1862, the refugees began to appear in increasing numbers at Cassville, Missouri.

The First Arkansas Union Cavalry

Captain Marcus LaRue Harrison of the 36th Illinois Infantry was the quartermaster of the Federal garrison at Cassville. He liked these Arkansas men who had given up their homes for the Union. When he was authorized to raise a company for the 6th Missouri Cavalry, he decided to fill the enlisted ranks with them. When the quota was quickly filled, the idea emerged to form an entire regiment from this pool of manpower.

Harrison was the son of a New York Presbyterian minister. He went to Yale to study theology, but instead became a teacher in southern Illinois and eventually became a civil engineer in the railroad business. For a time he worked in Chicago for the Burlington & Quincy Railroad Company, and then as the "Master of Buildings and Car Repairs" on the Burlington & Missouri River Railroad. Later, he moved to Iowa.

Many Yankee officers did not agree with the 32-year-old Harrison's

Colonel
Marcus LaRue Harrison
Shiloh Museum.

enthusiasm for these men. A regiment of Union men from seceded Arkansas was, many Northerners thought, just a plain bad idea. The belief was commonly held that such men had divided loyalties and in the final analysis would not fight as soldiers must. "There are officers in the army," Bishop wrote later, "who knowingly shook their heads at the project, and prophesied nothing but failure."[18]

Yet others felt quite differently. Brigadier General Egbert B. Brown, the commander of the District of Southwest Missouri, and Arkansas Federal Military Governor John S. Phelps were instrumental in getting the regiment approved in Washington, D.C. Finally, on June 16, 1862, the War Department issued a special order to Captain Harrison, stating that "the Secretary of War hereby authorizes you to raise a regiment of cavalry from the loyal men of Arkansas, to be completed by the 20th of July, and to be mustered into service, clothed, mounted, and armed at Springfield, Missouri, by the United States government. The regiment will be mustered into service for three years or the war...."[19]

Meanwhile, Arkansas men kept coming into Union lines. On May 10th, eleven men led by Thomas J. Gilstrap came to the Federal picket line. Four days later was the arrival of Thomas Wilhite and his thirty. Then on June 20th, a hundred and fifty men from Washington County rode in under Thomas J. Hunt. There was no doubting that Arkansas Unionism was on the rise.

Staffing the new regiment was a big project. Officers and men came from many sources and backgrounds. Colonel Harrison did not go into the regiment alone, but took with him his family and friends. His younger brother Elizur, age 22, entered as a private but soon found himself a Lieutenant. Also, the Colonel's son, Edward M. Harrison, enlisted. Although only 12 years old, he was registered as 18 and made a bugler. Two of Harrison's compatriots from the 36th Illinois Infantry, Sergeant James J. Johnson and Quartermaster Sergeant James Roseman, were brought on board as First Lieutenants of Companies A and M, respectively.

The position of second in command of the regiment fell to 30-year-old Albert W. Bishop, a Yale-educated lawyer. A native of New York, he started a family and law practice there. When his wife died in 1860, he moved to LaCrosse, Wisconsin. Upon the outbreak of the war he became a Lieutenant in an artillery battery, then moved on to be a Captain in the 2nd Wisconsin Cavalry. Bishop brought with him that regiment's chief trumpeter, Albert Pearson, who was made Second Lieutenant of Company K in the new Arkansas regiment, and hospital steward Frank Strong, who was made a Second Lieutenant and regimental adjutant.

Thomas J. Hunt.
Author's Collection.

Other commissioned officers for the new regiment came from two sources. First, authorizations to raise troops were given to local men from southwest Missouri and northwest Arkansas who were thought sufficiently eminent in their home districts to recruit a number of men. Thomas J. Hunt, a 23-year-old school teacher who was born and raised in Washington County, brought in a good many men and was made Captain of Company B. Sergeant John I. Worthington of the 6th Kansas Cavalry, formerly of Carroll County, Arkansas and presently of Granby, Missouri, was authorized to raise a company. Upon doing so, he became Captain of Company H. Civilian Charles Galloway of Barry County, Missouri, brought in many recruits and was appointed Captain of Company E. Thomas Wilhite of Washington County, Arkansas, was rewarded with the rank of Lieutenant for his recruitment success.

A second source of commissioned officers was the lower ranks of existing Federal units. The formation of the First Arkansas Union Cavalry created a great opportunity for promotion for many enlisted men of other regiments. Transferees included Private James H. Wilson of the 37th Illinois Infantry (promoted to First Lieutenant of Company D), Corporal Jacob J. Reel of the 4th Iowa Infantry (First Lieutenant of Company I), Corporal Henry W. Gildemeister of the 1st Missouri Cavalry (Second Lieutenant of Company I), Private John Bonine of the 4th Iowa Infantry (Captain of Company L), and Private Joseph S. Robb of the 4th Iowa Infantry (First Lieutenant of Company L). There were others as well.

The recruiting of enlisted men went on in two areas - in Missouri from Missourians and the walk-in refugees fleeing from Arkansas, and in Arkansas among Union men still at home. W.C. Peerson of Company B illustrated the former situation. He wrote the near the end of the war of his motives for enlisting:

> *I was borned in Washington County [Arkansas] on the 23rd of March 1844. I lived with my Parance until about the age of 18. In the year 1862 (the most of that time we lived in the State of Arkansas) I was forced by the Rebels to leave my home and take sides with them or go North to the Federal Army and I took it my choice to go North - myself, my Father, other relations and acquaintances 8 in number.*
> *Started on the morning of the 1st of December 1862 for the Federal Army, of which the nearest point was Elk Horn in Benton County, a distance of 35 miles. There was some Rebels between where we started from and the Federal Army although we had no trouble on the route. We reached the Army on the night of the same day, where we remained until the 3rd of December when myself and friend Campbell joined Co. B, 1st Arkansas Cavalry Volunteers on the morning of the 4th.*

Other recruiting efforts went on secretly in Confederate territory. On June 27, 1862, General Brown reported "Carroll, Madison, Benton, and Washington Counties have been thoroughly scouted..., the expedition bringing in about 100 recruits for the First Arkansas Regiment.

Northwest Arkansas is reported loyal, and its permanent occupation would demonstrate it."

The primary means of recruitment was for word of mouth of a clandestine meeting in a secluded location to be whispered among known Unionist sympathizers. There was danger both in the telling of the meeting and in the attending of it. "Recruiting in Arkansas for the Union Army was at that time a perilous undertaking," wrote Lt. Col. Bishop. "Loyal men avowed their principles at the hazard of life, and the greatest difficulty was in getting recruits to the rendezvous of the regiment for which enlistments were being made."

"From the enlistment of its first man," Bishop continued, "to the mustering of the twelfth company, the camp of the regiment was a continuous story of wrongs and outrages, and old men and boys, women and children, were subsisted by the Government, whilst husbands and brothers were preparing for the avenging strife. Singly and in groups they came to Springfield. Weary and sore, they stood up to be 'sworn in,' many infirm of limb, but firm of purpose, and thus arose the regiment."

Humiliation and Disaster
at the Battle of Prairie Grove

On October 18, 1862, the First Arkansas Union Cavalry was ordered to establish a forward post at Elkhorn Tavern, Arkansas. That was the very site of the Federal victory at the "Gettysburg of the West," the battle of Pea Ridge, which had taken place the previous March. Undoubtedly a wave of excitement swept through the men upon this announcement. Going back to reclaim Arkansas for the Union was exactly what most of them had enlisted for.

While the third battalion of the regiment was still organizing in Missouri, the first two battalions were at Elkhorn. The assigned duty of the regiment at that advanced Union outpost was a difficult one. A letter from Union General John M. Schofield was received which set forth this virtually impossible order:

> *That no misapprehension may exist, this is to inform you that your forces are expected to continually scout and scour all the country within your reach. One-*

half of the command may be on distant scouts all the time; the other portion should be constantly employed in your immediate neighborhood. No part of your forces should be idle at any time. You are expected to rid all the country within your reach of all small bands, guerrillas, provost guards, etc., etc. Your forces should continually harass the enemy by driving in pickets and skirmishing with advanced guards and detached parties, capturing forage trains and commissary wagons.

No limit is placed upon the country through which you may act, but you are expected to go wherever you can, without necessarily jeopardizing your command. You are to relieve the Union people and punish the treasonable. Unfailing activity and the utmost vigilance are demanded at your hands. One large party, consisting of about one-half your men, should be pushed near the enemy's lines and kept out all the time, capturing pickets, etc., and you may even go in the rear of the enemy's forces, and do them all the damage you possibly can. Feel the enemy often, and communicate all information you may obtain. This force should be relieved by the other half after a scout of five or six days.[20]

Lt. Col. Bishop, who was in command at Elkhorn while Colonel Harrison was on detached duty in Missouri, was incredulous. "All this," he marveled, "was expected from two battalions of cavalry, who had never been one hour in a camp of instruction; and though now in the service from eight to nine months...had...been only partially clothed - there was not an overcoat in the line - and has never been paid.... But the men knew the country where they were operating. They were in their native hills again, and were active and zealous in the efforts to support that Government, loyalty to which had caused them so much suffering."

Meanwhile, larger events were occurring. Union General James G. Blunt and Confederate General Thomas C. Hindman were both spoiling for a confrontation, and as December came on, it was clear that the moment of decision was close at hand. The First Arkansas Union Cavalry was about to have its first real battle.

On Saturday morning, December 6, 1862, Colonel Harrison left Elkhorn Tavern with all three battalions as part of the southward movement of General Francis J. Herron's army, which was hurriedly moving to join with General Blunt before Hindman's Confederate forces could strike. The next evening the First Arkansas arrived at the Illinois River southwest of Fayetteville and encamped. Colonel Harrison sent a message to General Blunt, who was at Cane Hill, just a few miles away, stating that his men and horses were too tired to proceed further, and that he did not think he could move them before Monday morning.

Blunt was furious with this message. It was Saturday night, Confederates were about to attack him, and when within only a few miles of helping them, Harrison sent a note like this. Harrison's men had come a shorter distance on horses than General Herron's men had come on foot, and the infantry was still coming. In his report on the Battle of Prairie Grove, Blunt justifiably complained:

> *About 9 p.m. of the 6th, I received a note from Colonel Harrison, of the First Arkansas Union Cavalry, who had been ordered down from Elkhorn at the same time that General Herron started from Wilson's Creek, informing me that he had arrived at Illinois Creek, 8 miles north of Cane Hill, with five hundred men, and that his horses and men were so tired that he did not think he could move farther until Monday, the 8th. Whether his regard for the Sabbath or the fear of getting into a fight prompted him to make such a report to me, I am unable to say; but, judging from his movements that he was not a man upon whom to place much reliance on the battle-field, I ordered him to proceed by daybreak to Rhea's Mills, to guard the transportation and supply trains....*[21]

Unfortunately, this note by Harrison remains a mystery. In the records available today, no explanation is made by him on this subject. Bishop in his writings did not comment either, which may have been because he was not present (he remained in command at Elkhorn Tavern), or perhaps because he did not wish to criticize the commander under whom he served at the time of writing and publishing. Clearly, however, it was a personal low point for Harrison personally. The regiment as a whole was going to hit its own low the following day.

On the early morning of December 7, 1862, the Confederates put into motion their bold plan to first strike and defeat Herron, then turn on and destroy Blunt. Between Blunt and Herron were a few regiments of Union cavalry, which were the vanguard of Herron's army. The lead of these regiments was the 7th Missouri Union Cavalry, behind which was the 1st Arkansas Union Cavalry. The unsuspecting 7th Missouri was taking a rest from their fatiguing relief march when disaster struck. The last word from Blunt had been that the road to Cane Hill was clear of the enemy, and no contact was expected. It was felt safe enough that the 7th had the horses feeding "with bridles off and girths loosened." The 1st Arkansas, as ordered, was working its way south to Rhea's Mill to guard the supply trains.

It was at this point that General Hindman's Confederate cavalry came crashing through the Missourians' lines. The Federals were utterly unprepared. They put up a very brief resistance, but it was soon overwhelmed. "The command...retreated in every direction," reported Captain M. H. Brawner of the 7th Missouri, "quite a number running into the rebel lines, being killed or captured."

The Missouri Federals fled in complete disorder back up the road to Fayetteville and directly into the First Arkansas Union Cavalry. Panic in turn seized the Arkansans, and they turned and fled in chaos. "This was caused partly by the overpowering number of the enemy," reported the Regimental Return for December 1862, "and partly by the disorderly retreat of the Federal cavalry in advance of our Regiment." Colonel Harrison wrote afterward that he was left behind "in the extreme rear of my men who had all left me."[22] The twenty regimental wagons were abandoned, and only the one bearing the regimental flag was able to be brought out safely.

"The fight grows intensely interesting," reported Confederate General Jo Shelby, commander of the onrushing Rebels, "and my men, feeling the inspiration of the scene, dash on and on, taking prisoners, capturing guns, colors, horses, mules, and every form and variety of clothing, left in the desperate flight of the terror [stricken] enemy."

The retreating Federals fled headlong for some four to five miles.

They went right through the 1st Missouri Union Cavalry, which tried to stem the tide. Captain Amos L. Burrows reported:

> [W]e saw a large body of the First Arkansas and Seventh Missouri Cavalry on the retreat. We undertook to stop them, and, finding it being of no use, Major Hubbard ordered the fence to be thrown down on the lefthand side of the road, and drew up in line of battle in the wheat-field, and instructed the First Arkansas and Seventh Missouri to form in our rear. They partially did so, but, having several shots fired at our line by the enemy, they broke and fled.[23]

Benjamin F. McIntyre of the 19th Iowa Infantry was with General Herron's main army. As his regiment hurried to the front, he wrote that "the 1st Arkansas Cavalry came rushing by us on their horses completely panic-stricken - many without hats or coats, spurring their animals to the utmost speed. Our force had no influence on them to return and pell mell completely frightened they rushed on by us. This was not encouraging to raw troops and this display on the eve of a fight by Union Cavalry was no such a display of gallantry as we had anticipated and I heard the word Coward fall from more than one boy's lip...."[24]

General Francis J. Herron.
Library of Congress.

Finally, the flight reached General Herron. "They came back on me 6 miles south of Fayetteville, at 7 a.m.," he wrote, "closely pursued by at least 3,000 cavalry. It was with the very greatest difficulty that we got them [the retreating Federal cavalry] checked, and prevented a general stampede of the battery horses; but after some hard talking, and my finally shooting one cowardly whelp off his horse, they halted."[25]

The First Arkansas was diverted into a field by the Walnut Grove Church to re-group. "After I had gained the advance portions of my men," Colonel Harrison wrote, "I halted before them with revolver drawn in a number of instances and with the assistance of several of my officers partially stayed the stampede."

The story of the First Arkansas Union Cavalry at the Battle of Prairie Grove was called by Lt. Col. Bishop "a chapter of accidents." Colonel Harrison called it a "counter-march." Several officers of the regiment in a joint letter called it a "retrograde movement." What it really was, was an embarrassing, unmitigated fiasco. Four men were killed, four wounded, and forty-seven captured or missing. Commanding generals believed the colonel to be unreliable. Orders had been perceived to be disobeyed. The regiment had fled in panic. General Blunt in his after-battle report not only contemptuously chastised Harrison's evening note that his regiment was too tired to move on, but then condemned his getting caught in the Confederate's early morning attack. "Had he, instead of making unnecessary delay, promptly obeyed that order [to move to Rhea's Mill], he would not have had a portion of his command and transportation captured."

Perhaps the criticism had been right - that loyal Arkansas men would not, could not, fight.

Following the battle of Prairie Grove, a Federal post was established at Fayetteville. Notwithstanding the disaster and rebuke for his and the regiment's conduct in the battle, Colonel Harrison was placed in command. This was surely because the men of the regiment were from the area of Fayetteville, and knew the town and the people there and in the surrounding countryside.

The poor reputation of the First Arkansas continued. Colonel William A. Phillips was in command of the district, which consisted of his own command on the border of the Indian (Oklahoma) Territory and of Harrison's post in Fayetteville. On March 27th he reported to Major General Samuel R. Curtis, the commander of the Department of the Missouri. "The enemy still hold Clarksville, although no point on this side of the river above it. They hold Fort Smith, from which I could very easily dispossess them, but have been embarrassed by the Arkansas command at Fayetteville.... The report of the Arkansas command, Colonel Harrison, shows several things I have been trying to correct. I believe Colonel Harrison does the best he can with it, and

I hope that more rigid discipline may be gradually introduced."

Without any doubt, Harrison did have disciplinary problems with the First Arkansas Union Cavalry. One of the primary causes was that the men were stationed at a post very near their own homes. Soldiers were coming and going to their families without regard to military necessity or permission.

Another factor was that these men were from backwoods hill country where military discipline was about the last thing they cared about. An interesting order was put out by Colonel Harrison to help deal with their near-lawlessness:

> *Head Quarters, Post of Fayetteville Ark.*
> *March 11/63*
> *General Order No. 10*
>
> *The firing of arms at any hour of the day or night without written permission from a company Commander..., the cutting or injuring of shade trees, shrubbery or fruit trees; the demolishing of fences or the trespassing in any manner upon private property...will be severely punished. All officers and guards are ordered to make it their especial duty to arrest and place in confinement any person or persons found guilty of violating this order....*
>
> *By Order of Col. M. La Rue Harrison*
> *Commanding Post*

Such was the story of the First Arkansas Union Cavalry.

In this assignment at Fayetteville, the regiment nonetheless performed the hard duty of the Union cavalry in a remote outpost surrounded by enemy sympathizers. The constant wear and tear of the demanding service there was reflected in the regimental return of February 1863. "The Regt is cept [sic] constantly on duty," it recorded. "Scouts are sent almost daily.... Discipline is tolerably good, for drilling we have no time the men are always off on duty."

This was confirmed by the company return of Company M for the same month: "[T]he company have furnished details for escorts, messengers, mail carriers and been so constantly on duty that but few horses are left in the company, and owing to scarcity of forage, those now remaining are nearly all unserviceable."[26] This was the occupation of the First Arkansas Cavalry when the Battle of Fayetteville occurred.

The First Arkansas Union Infantry

After the ranks of the First Arkansas Union Cavalry were filled, Arkansas men still continued to pour into Federal lines. It was decided to organize an Arkansas Union infantry regiment. Thirty-year-old Dr. James M. Johnson of Huntsville, Madison County, was selected as the new Colonel of the First Arkansas Union Infantry. He was a firm Unionist and friend of Isaac Murphy, the last anti-secession delegate. Like so many other Union men, they fled Arkansas for their lives, leaving together in the spring of 1862.

The enlisted men were recruited from Madison, Washington, Newton, Benton, Searcy and Crawford counties in Arkansas. Young Captain Elhanon J. Searle from Company M of the 10th Illinois Infantry was appointed as the Lieutenant Colonel of the regiment. The First Arkansas Union Infantry was officially mustered into the service of the United States on March 25, 1863 in Fayetteville. It was by no means prepared in training or equipment to take a significant role in the Battle of Fayetteville.

A third Arkansas unit was also in the process of being formed at the time of the Battle of Fayetteville. The First Arkansas Union Light Artillery was authorized and recruiting, but was not yet officially mustered or armed. It was, in fact, entirely absent to Missouri on the day of the attack.

The Approach of Battle

"I hope you will move on Fayetteville," Confederate General William Steele wrote to General Cabell on March 12, 1863. "My information is that there are only about a thousand men there, and no cannon."[27] This intelligence was accurate.

Furthermore, in terms of defending the place, Fayetteville was in an inherently exposed position. It was fifty miles beyond the Federal base at Cassville, Missouri, and could not be supplied by either train or riverboat. The deteriorating condition of the horses rendered the outpost even more isolated, as it could not maintain an effective mounted reconnaissance of the surrounding countryside. The post could neither be quickly reinforced nor rapidly withdrawn. It was an inviting target for the Confederates to attack. Its greatest defense was the corresponding weakness of the Confederates in northwest Arkansas.

Upon receiving the further but erroneous information that the Federal outpost at Fayetteville was preparing for withdrawal, General Cabell decided that the opportune moment for a "dash" had indeed arrived. Cabell's' reasoning for striking now was given in his report after the battle:

> *Knowing that our good citizens had burdens imposed on them by the Federal troops too grievous to be borne much longer; that it was necessary for me to visit that section of the country, and having been appealed to by citizens, both male and female, to give them assistance, I determined that I would strike there the very first time that I saw the least hope, whether I succeeded in taking the place or not.*
> *As soon... as I learned that Phillips was moving around with his Indian brigade to flank General Steele, and, having consulted with General Steele, who agreed with me (and desired that a dash should be made at Fayetteville, if nothing more) that it was necessary, and having heard that they were getting their wagons ready (which proved to be false) to reinforce Phillips,*

> *besides being without forage (nothing to feed my horses) I determined to make a bold dash at that den of thieves and if possible to take it.*[28]

Assembling his brigade at Ozark, about 75 miles by road to the south located along the Arkansas River, Cabell made ready for the assault. The Confederate expedition against Fayetteville numbered about 900 men. Except for the three companies of Colonel Hill's battalion of cavalry which were left behind in Ozark, the entire brigade was brought along.

General Steele later described the attacking force. "General Cabell's brigade," he wrote, "having been assigned to my command, would convey the idea of a respectable force, which is an erroneous idea. Monroe's and Carroll's regiments, both weak, are all that have ever been here. The balance consists of companies and battalions scattered through the country...and...it is a matter of doubt if any great number of them can be brought together."[29]

The Confederate position concerning forage for horses was desperate on the day Cabell's cavalry left for Fayetteville. "I had all my troops together the day I left for Fayetteville...," he wrote afterward. "I did not have the day I left half forage for my command, and was compelled to moved somewhere. I considered it necessary to find out by going myself to see if I could find any subsistence north of the [Boston] mountains...."

Colonel Harrison was not in a prime condition to receive this Confederate attack. He had under his command at the post of Fayetteville the First Arkansas Union Cavalry, the First Arkansas Union Infantry and the First Arkansas Union Light Artillery. The apparent strength, however, was not actually present. In March Colonel Phillips, commander of the District of Northwest Arkansas, reported, "At the post Fayetteville is the First Arkansas Cavalry, in poor condition; First Arkansas Infantry, 400 present, absent, sick etc., and a battery (the men without guns), the latter two forces being of no consequence at present. The whole force is not there; is not as great as it appears on paper..."

Harrison gave his own assessment of the strength of his command some two weeks before the battle:

Headquarters Post, Fayetteville, Ark., April 1, 1863
Maj. Gen. Samuel R. Curtis, Commanding

General...
      1st. The state of my command. The First Arkansas Cavalry numbers an aggregate of 1,032 men; probably when all are at the post they may number 850 effective men. They have 154 serviceable horses and 65 unserviceable, all told. The regiment has not received any clothing for three months, and only a very small supply since November, so that a large part of the men are in a destitute condition.
      The First Arkansas Infantry will number in a few days an aggregate of 830 men; probably 700 of them effective. They are totally without transportation, clothing or tents, or equipments of any kind, except the arms picked up on the Prairie Grove battle-ground, which are of all patterns and calibers. The destitution of clothing is very great, and much suffering and sickness prevails on account of it; besides it would be a ruinous policy to place this undrilled, barefooted, butternut regiment in the field to be mixed up and cut in pieces by rebels in the same dress.
      The First Arkansas Light Artillery numbers 110 men, who are destitute of clothing, and have never received their guns. Of course, nothing can be expected of them...
      M. La Rue Harrison
      Colonel First Arkansas Volunteers Cavalry
      Commanding Post[30]

Rumors were rife that the Rebels would strike at Fayetteville. On April 2nd, Phillips ordered to Harrison with orders to be prepared: "Call in the command, and keep it at the post. Throw up earthworks as speedily as possible. Defend yourselves as you see fit, but lose no time.... Put your men in effective shape. Make their position strong. Exert yourself so as not to embarrass me.... I...have confidence in your judgment with *your peculiar command.*"[31]

The old Yankee suspicion about Federal Arkansas troops is apparent in this "peculiar command" comment. Loyal Arkansas soldiers still had a point to prove to their fellow countrymen - that they could and would fight. Their opportunity to do so was fast approaching.

On April 11th Harrison wrote to Major General Herron that reports had reached him of "three pieces of artillery shipped to...Ozark, for the ostensible purpose of attacking Fayetteville," and of a plan to "attack Fayetteville with the cavalry and some artillery" with "a raid...intended next week...." This intelligence would prove to be exactly accurate. The following day, Phillips told Harrison that if he was hard pressed he should retreat toward Phillip's command.

The regimental muster roll dated April 12, 1863 stated: "This regiment since last muster has been constantly engaged in scouting, guarding forage and other trains and in doing the general duty of an outpost on the frontier." The April 17th Morning Report of Company F noted, "All company horses turned out as unserviceable."

The Post of Fayetteville Field Return dated April 20, 1863, shows that of the 36 officers in the First Arkansas Union Infantry, three to five were gone. Colonel James Johnson was gone to St. Louis as a witness before a military commission. Captain James R. Vanderpool of Company C was out on a scout with fifty men in Newton County, and Captain Abial Stephens led another scout. One officer was sick. Part of Company E was absent. First Lieutenant Aaron C. Terry commanded only a portion of Company I.

Of the approximately 838 enlisted men belonging to the First Infantry at the time, 41 were sick, 46 were absent without leave, 124 were absent with leave, and 142 were out on patrol or otherwise on detached service. This left a total of about 485 men and 32 officers present.

Those that were present were not properly armed. The regimental return of Company I reported that only a "portion of the Company were engaged in the Fayetteville fight. But a small number of the Company were efficiently armed."

Also, a significant part of the First Arkansas Union Cavalry was not present. Of its 45 officers, twelve were gone. Eight were on detached service on scouts. The other four were absent with leave, absent without leave, "in the hands of civil authorities" (whatever that meant), and under arrest.

Nearly half of Company D was gone on a foot scout under Lieutenant Wilhite in the Boston Mountains. Company F had only 16 men present under Sergeant John Dienst, with the remainder absent with Captain Richard Wimpy on escort duty to Cassville, Missouri. Company M had only 13 men present as the rest were under Lieutenant John Turman on forage duty. Company H had no officers present and was under the command of 2nd Master Sergeant John G. Black.

Of the cavalry enlisted men, 976 were on the books at this time, but a considerable number of them were away on the day of battle. This included 62 sick, 23 under arrest or confinement, 20 absent with leave, 84 absent without leave, and 223 on detached service. That left about 554 men available for the battle.

In addition, on the day of the battle one assistant surgeon and eleven privates were on duty at the hospital. All told, then, approximately 1,116 Union officers and men were available in the defense of Fayetteville on April 18, 1863.[32]

Also present in the area, by chance, was a train escort newly arrived from Springfield, Missouri. It had elements from the 2nd Kansas, 11th Kansas, 27th Wisconsin and 37th Illinois. These latter units played no part in the battle except to be taken as prisoners by the Confederates. They were available, however, if ultimately needed.

All things considered, some nine hundred Confederates were attacking about eleven hundred Federals. In a one-to-one attack, it was obviously going to be very tough going for the Confederates, but, as Cabell stated afterward, he had mistakenly expected the Union troops to be in the act of vacating the town. If this had been true, then the odds would likely have shifted in his favor. But it was not true. As it was, Cabell still had two things in his favor - surprise and artillery. If he was to win this battle, he would need to make good use of both.

Cabell's Brigade left Ozark at about three in the morning on Thursday, April 16th, with three days rations and a full supply of ammunition. The Rebels rode until noon, then rested until sunset. In the dark, they made their way toward Fayetteville up the Frog Bayou Road, which is roughly parallel to the modern US highway 71 today. Approaching town from the West Fork of the White River; they crossed the David Walker farm and went up to the south slope of East Mountain and into the ravine that sloped away from the Federal position on the east side of town toward the mountain.

The Battle of Fayetteville

Although the Union defenders at Fayetteville were expecting a Confederate attack to occur sometime in the near future, they were not in immediate expectation of it on the evening before it actually occurred. Lieutenant Joseph S. Robb of Company B of the First Union Cavalry returned to camp on Friday the 17th, reporting that his scout in the direction of Ozark revealed "no apparent preparations of the enemy to move in this direction." Colonel Harrison took the report at face value. Lacking the horses to maintain a constant mounted patrol, he decided to wait until the next day to send out another scout. Lt. Robb, by a very considerable margin, had missed the very purpose of his scout, returning to his post with a Confederate army only hours behind him.

Throughout Friday night, April 17th-18th, the Confederates closed in on Fayetteville. Moving along the West Fork of the White River, they advanced northward and then eastward toward town. Two people were of particular benefit in guiding the Rebels that night. One was Sergeant Mathew W. Sumner of Company A of the First Union Infantry, who had deserted from Fayetteville just two days before and gone over to the enemy.[33] Sumner, who must have known the most recent Federal dispositions around town, was likely a very great aid to General Cabell. A second guide was a young boy by the name of R.J. Wilson, a local resident of about 12 years of age who also could give good information on Union troop placement and local geography.

Moving through the night, First Lieutenant James A. Ferguson of Company F of Carroll's Cavalry, came across a farm house full of light, people and music. This was at West Fork on the White River east of Fayetteville, nine miles from town. Finding no guards, Ferguson's men surrounded the place. Inside, Union soldiers, absent without leave (AWOL) from the post, were kicking up their heels with some young ladies at a dance.

The fun came to a quick halt when Lt. Ferguson stuck his head in the door and called upon the party goers to surrender. The bluecoats scrambled, but outnumbered and surrounded, they wisely gave up the struggle very quickly. Union Lieutenant Gustavus Hottanhour was the only officer present. Nineteen days earlier he had been promoted from First Sergeant to Second Lieutenant, and now he started off his career as an officer with this incident. He and eight enlisted men, in this ignominious fashion, found themselves to be prisoners of war. They were detained until after the battle and then paroled. In praising the most meritorious officers and men after the battle, General Cabell wrote that Lt. Ferguson was among those that "deserve to be particularly mentioned."

Moving in closer to Fayetteville, a few minutes after sunrise the Confederates encountered the dismounted Union picket just east of town. These Federal troops received a much rougher handling than those at the dance. It was just about sunrise now, and it was time for the attackers to make their move.

On April 18, 1863, the sun rose at approximately 5:40 a.m. No one complained of the weather that day, so it must have been relatively fair and not a detriment to either side. Spring weather can vary widely in Fayetteville, and temperatures in the 1800s were somewhat cooler than now, but modern averages on that day are 49 degrees for a low and 72 degrees for a high. It is probably reasonable to say that the temperature was about 50 when the battle began, and rose to the low 60s by its end.

The Rebels quickly overran the picket post, but shots were fired before it was all over. Privates Jonas Riddle of Company A and Lucien Amos of Company H were killed. Private George W. Russell of Company G was captured, identified as having been a Federal spy in the past, and hanged in execution. Others were also captured.

Though killed and captured, the Union guards had nonetheless successfully performed their mission. "The firing of the picket had alarmed the command," Colonel Harrison wrote afterward. General Cabell's element of surprise, so carefully preserved up to this moment, now began rapidly to dissipate.

With the enemy alarmed, the Confederates moved as quickly as possible to get into a position to launch their attack. This was not easy. The men had to be brought up and dispositions completed. The two pieces of artillery were in the rear and had to be brought forward, taken up the south side of East Mountain, and made ready. The time and noise in this effort ended what little advantage of surprise still lingered.

The Union troops were now reacting to what was happening. "The soldiers were still in bed when the first alarm was given," wrote Sarah Yeater, who lived in the Baxter house that day. The commotion awakened Federal Lieutenant Elizur B. Harrison, the younger brother of the post commander. "About daybreak on the morning of April 18th," he recalled many years later, "I was awakened by an unusual noise, and hastily dressing, I opened the east door of my room [in the Baxter house] and to my consternation saw near the back of the lot a column of Confederate cavalry. It is needless to say that I hurriedly shut the door and made my getaway through the front door down to College Avenue." which was then known as Cassville Road.

Once out of the house, Lt. Harrison found his brother, the Colonel, coming away from Headquarters House across the street from the Baxter home. Joining him, they hurried together over to the camp of the First Union Cavalry, which lay north up the Cassville Road. "We found the men getting into their clothing, gathering arms and ammunition," the Lieutenant said, "while the officers were getting the men into order."

As previously mentioned, the Federals were expecting an attack but not on this particular morning. Colonel Harrison was, as Dr. Seymour Carpenter described, "partially taken by surprise." Nonetheless, he responded quickly and "had time to get his men in position before the attack was made." Harrison set both the First Cavalry and the First Infantry into motion. He did not know how many men the enemy had, what their troops dispositions were, or what their intentions were. All he could do was react, and he did it well, setting up a main defensive line with protections on the flanks and a reserve. He ordered "the First Sergeants to personally see that their companies were supplied with ammunition." He ordered 25-year-old Lt. Col. Elhanon J. Searle, who had formed the Infantry on its parade ground, and who commanded in the absence of Colonel James M. Johnson, to slowly withdraw his exposed regiment on the east end of town toward the position of the cavalry.

The First Union Cavalry, on foot throughout the battle, was ordered into position to receive the attack. The third battalion (consisting of only two companies) was placed on the right, under the command of Major Ezra Fitch. The second battalion under Lt. Col. Albert W. Bishop and Major Thomas J. Hunt was put on the left. The center, commanded personally by Colonel Harrison, was the strongest point. It consisted of four companies of the First Cavalry and three companies (A, F and H) of the First Infantry. "Fearing that, not being uniformed, they might be mistaken for the enemy," Harrison put the rest of the infantry (seven companies) in reserve in a sheltered position to the rear under the command of Lt. Col. Searle. Captain Rowen E. M. Mack of Company G of the Cavalry (mostly men from Carroll County, Arkansas) was ordered "to reconnoiter on the far right to prevent a flank movement in that direction." Captain Hugo C. Botefuhr and Company C were placed in reserve.[34]

One of the Federal officers moving to his post was Lt. Alfred Hutchison of Company F, who was under arrest and awaiting the decision of a court martial. He had tried to bribe one of the regimental surgeons into giving an enlisted man in his company a medical discharge. All disputes were forgotten in the urgency of the moment, and he "repaired promptly [to] his command and was with it during the engagement."

During the time of deploying, General Cabell was surely already beginning to feel the first of the many frustrations to afflict him that day. It was already apparent that the Federals were not getting their wagons ready to pull out of town, as he had been told. "Our friends," he lamented afterward, "are all too anxious to rid the country of their presence to state things as they really are." Cabell also found what he estimated to be a good many more Union soldiers that he anticipated, "notwithstanding all previous reports from persons living in Fayetteville to the contrary." Now the advantage of his hoped-for surprise was just about entirely gone, and his artillery was still not ready.

About six o'clock the Confederates made their initial move toward Fayetteville, charging on horseback "with wild and deafening shouts" up the ravine on the east side of town toward the Federal Headquarters and the Baxter house. This attack was under the immediate command of Colonel John Scott, and included Carroll's Cavalry under Lt. Col. Lee L. Thomson and Dorsey's Missouri squadron, which was commanded by Colonel Dorsey in person.

This "dashing charge," wrote Cabell, "drove the enemy to their pits and to the houses, where they rallied and poured in a dreadful fire with their long-range guns." John M. Harrell, who later served as a Colonel in the brigade but was not present at this battle, wrote after the war, "In the streets Cabell's men met with effectual resistance from the windows, doorways and corners of the houses...." Henry G. Orr of Parson's Texas Cavalry wrote that the Confederate attack sent "the enemy seeking protection in and behind houses and some in rifle pits." In the Civil War rifle pits were not foxholes in which the soldier would stand, but rather shallow trenches in which he would lay down and fire, with the dirt from the hole piled in front of him for protection.

In making the attack, Benjamin H. Ashcom of Dorsey's Squadron was almost simultaneously wounded three times. In defending the attack, First Sergeant William M. Burrow of Company E of the First Arkansas Union Cavalry fell badly wounded. "As his comrades were bearing him from the field, he begged them to 'lay him down and go to fighting.'" Burrow died of his wound two weeks later.[35]

While the Confederates lamented about rifle pits and doorways, Harrison had a different complaint. "During the whole action," he said, "the occupied ground covered with timber and brush, while my command were in the streets and open fields." Of the two situations, the Federals clearly had the better one, and the Rebels the greater cause for complaint.

*Confederate Artillery
at the Battle of Fayetteville.
DeeDee Lamb.*

As the Confederates neared Federal Headquarters and the Baxter house, and the Federals occupied their rifle pits and houses, Cabell stopped his men and waited for the artillery to get into place. The cannon fire did not inflict great casualties, but it did devastate the enemy's morale. There was no Union artillery to return fire. Meanwhile, dismounted Rebel cavalry was sent against the Baxter House and Federal Headquarters.

Within thirty minutes of the initial attack, Lieutenant Hughey had his two-gun section of artillery in place and ready for action. Located on the Confederate right above a spring on East Mountain, near the home of Confederate Colonel Thomas Gunter, the guns were protected by visibility-obscuring brush and by a skirmish line under Captain Joseph P. Weir's Company A of Parson's 12th Texas Cavalry. Weir had been Captain of the company since the organization of the regiment in October 1861. It is probable that Captain Wilson J. Coggins' Company H of the Darnell's 18th Texas Cavalry was also at this point, all Texans being under the command of Lieutenant Colonel Sebron Noble. General Cabell set up battlefield headquarters just across the mountain road east of the artillery.

The artillery roared into action, firing canister and shell at the Union positions and causing a great deal of fear among many of the defenders. "Very soon after the first shots were fired a shell hit the jamb of the basement door," Sarah Yeater said of the Baxter house, "splintered it, knocked bricks from the chimney, broke a large kettle containing lye standing on the hearth, and rolled out of sight." According to Lt. Elizur Harrison, this cannonball was a fuse shell which was extinguished when it landed in the lye, preventing the explosion of the shell.

The Confederate guns focused on the camp of the First Union Cavalry. Cabell wrote that the artillery "did frightful execution in the enemy's camp, driving them out and completely scattering their cavalry for awhile." Colonel Harrison reported that the enemy guns "opened a sharp fire of canister and shells upon the camp..., doing some damage to tents and horses, but killing no men."[36]

"We were using the large brick smokehouse in the rear of the Van Horne lot as an arsenal," said Lt. Harrison, "and Jim Bell, a private in Company I, was stationed there to notify us if any attempt was made to capture our supplies." It was not long before a cannon "shot passed over the higher ground and in falling crashed through the body of Jim Bell at the arsenal.".

Not long after the guns commenced firing, a Federal soldier put his sights on the Confederate artillery commander. Lt. Hughey was wounded in the arm, but was nonetheless able to stay at his post and continue directing his men. "Captain Hughey deserves especial mention," Cabell reported, "for his bravery, skill and energy in the management of his two pieces of artillery."

One of the Texas troopers protecting the artillery was wounded in the fighting and went into town for attention. He went to a house located at the southwest corner of Lafayette and College, which admitted him to a bed for rest. It was not long before an artillery shell crashed into that very house of refuge and whizzed over his head. He lived to tell the tale.

Cabell reported that the artillery was "under a heavy fire from the enemy's sharpshooters during the whole fight." Although newly trained at the artillery instruction camp, and notwithstanding the fire, Hughey's men "with one or two exceptions did well." The effect of the intense artillery fire was to cause a considerable fear and demoralization of many of the green and untried Federal troops. Lt. Harrison stated that at its beginning the battle seemed destined to end in defeat for the Federals. Two men from Company A of the First Union Infantry, which was at the center of the defending line, broke under the stress of their first battle. Thirty-year-old Private Francis W. Cannon "run and concealed himself," and 18-year-old Private Gilbert C. Luper "was frightened and run when the rebels made their appearance." Corporal Thomas Bingham of Company E of the Infantry and Private Cyrus Barber of Company L of the Cavalry decided that it was time to flee. After the battle, Colonel Harrison reported that he had thirty-five men missing, "mostly stampeded toward Cassville during the engagement." He further noted that Lt. Col. Searle and Major Ham of the First Infantry "did good service in keeping their men in position [in reserve] and preventing them from being terrified by the artillery."[37]

But the fear was not confined to the enlisted men. First Lieutenant Crittenden C. Wells, quartermaster of the First Infantry, "ran away disgracefully to Cassville, Missouri." At the commencement of the bombardment about seven o'clock, Captain DeWitt C. Hopkins of Company I and First Lieutenant William L. Messenger of Company D, both of the First Cavalry, quickly lost hope under the cannon fire. In their minds they exaggerated the number of guns they faced. They

Map of the Battle of Fayetteville
(Prepared by the Author and Dan Quintans.)

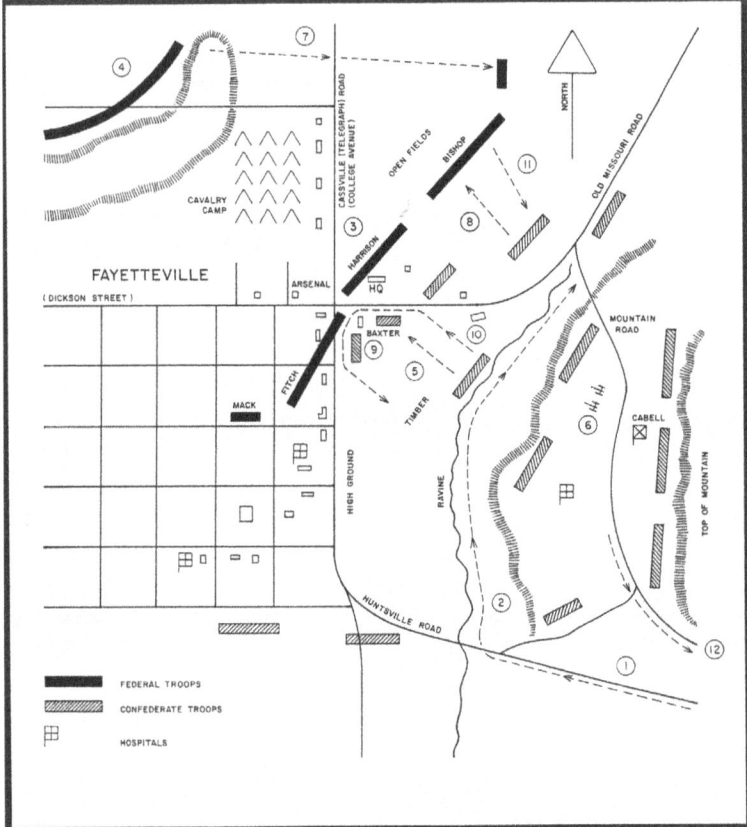

Battle Map Summary
(Numbered paragraphs correspond to numbers on the map.)

1. *5:40 a.m. The Federal pickets east of Fayetteville are quickly overrun by Confederates under Brigadier General William L. Cabell. Shots alert the Union defenders commanded by Colonel M. La Rue Harrison that an attack is imminent.*

2. *Using the ravine running north-south between East Mountain and the town, Confederates move closer to the Federal positions. Cabell sets up his field headquarters and a hospital on the hillside, with artillery to his front and reserves to his rear. Other Rebel troops occupy the*

southeastern part of Fayetteville.

 3. Harrison establishes a defensive line consisting mostly of the dismounted First Arkansas Cavalry. He commands the center, and places Lt. Col. Albert W. Bishop on the left and Major Ezra Fitch on the right.

 4. Most of the First Arkansas Union Infantry, which had no uniforms, is placed out of harm's way behind high ground to the rear. Three companies of Infantry, however, form part of Harrison's center line.

 5. 6:00 a.m. Colonel John Scott directs a Confederate mounted charge of Dorsey's Missouri Squadron (under Major Caleb Dorsey) and Carroll's First Arkansas Cavalry (under Lt. Col. Lee L. Thomson). The Union defenders disperse into houses, behind hedges and walls, and into rifle pits, and with their superior muskets bring the attack to a halt.

 6. 6:30 a.m. Lieutenant William M. Hughey's Confederate two-gun battery opens fire from East Mountain. It pours in a heavy fire nearly panicking the Federals, but they stand firm.

 7. Harrison sends two companies of the First Arkansas Union Infantry to protect the Federal left flank.

 8. Colonel James Monroe's First Arkansas Confederate Cavalry makes a dismounted attack in the open fields on the Union left. It fails to break the Federal line.

 9. There is fighting in and around Federal Headquarters and the Baxter house between dismounted men from Carroll's Cavalry and Harrison's Unionists. This is the area where the most casualties are inflicted. The Confederates capture the Baxter house, but the Federals hang on to Headquarters and its grounds.

 10. 9:00 a.m. Colonel Monroe leads a cavalry charge up the Old Missouri Road (Dickson Street), but runs into heavy defending musket and pistol fire from Federal defenders on the right and in front. The attack fails and the cavalrymen make a left turn and retreat.

 11. Lt. Robb of the First Arkansas Union Cavalry leads two companies to within musket range of the Confederate artillery and pours in a heavy fire. With ammunition exhausted, Lieutenant Hughey withdraws his artillery from the battle.

 12. The Confederate attack has failed. Desultory fighting continues on, but the battle is essentially over a little after 9:00 a.m. Cabell considers burning the town of Fayetteville as he pulls out, but decides not to do so because of Confederate families there. By noon, the Rebel army is gone.

went to Colonel Harrison, saying that six artillery pieces had been planted by the Confederates, and that the enemy was flanking them, presumably on the Federal left where the artillery was located.

"I replied," Harrison recalled, "'How can we retreat; they on horses and we on foot? Would you wish to be disgraced by a surrender?'" The spooked officers said they would not like to be so disgraced, they each vowed to "fight it out to the death," and returned to their commands. "And right well did they do their duty," Harrison said afterward. In response to the concern about flanking, Harrison directed Lt. Col. Searle of the First Union Infantry to send two companies (one of which was probably Company K) to the left of Lt. Col. Bishop's position to prevent being turned in that direction.

The demoralizing effect of the artillery, and the wavering it caused in the Union defense, was Confederate high tide at the Battle of Fayetteville.

Cabell now moved to test the resolve of the Union troops, finding out for himself whether Arkansas Union troops would fight or run. Monroe's Confederate cavalry dismounted and advanced on foot as infantry toward the Federal left. Lt. Col. Bishop, the Union commander in that section, called Monroe's assault a "bold advance." Union Captain William Johnson of Company M, who had just two months previously replaced the killed Captain Robert Travis, "had his right arm shattered while leading his men forward under a galling fire" in repelling Monroe's dismounted attack. This is the likely time that Private Davis Chyle and Corporal Doctor B. Norris, both of Johnson's Company, were wounded also. Out of 13 men of Company M present for the battle, three were casualties.[38]

Monroe's dismounted attack did not break the Federal defenses.

Although the strategic objective of the Confederates was the capture of Fayetteville, the primary tactical target was the Federal headquarters house in the center of the Union line. There was ongoing shooting and maneuvering, and the headquarters was repeatedly but unsuccessfully charged by dismounted Rebels. Bullet holes in interior doors of the house even today mark the ferocity of the struggle there that day.

The Tebbetts house was Federal headquarters in the Battle of Fayetteville. Author's Collection.

Bullet hole in the door at Headquarters House. Author's Collection.

Yankee firepower was overwhelming. General Cabell quickly learned to his chagrin that the Union weaponry was vastly superior to his own. The Federals were "well armed with Springfield and Whitney rifles," and "they poured in a dreadful fire with their longer range rifles."

The Arkadelphia guns used by the Rebels, Cabell lamented, were "no better than shotguns."[39] The rifles possessed by the Union soldiers were one-shot muskets with a rifled, or grooved barrel interior, which would put a spin on the musket ball, shooting it further and more accurately than the old smootbore guns. The Confederates had difficulty getting close enough to use their shotguns while always themselves within range of enemy rifles. The Rebels were at a serious disadvantage with their inefficient arms, and, in the final analysis, the disadvantage could not be overcome. Cabell claimed that the Federal rifles could shoot as far as his artillery, which seems an exaggeration, but may have been a correct statement in the close confines in which this battle took place.

With the Baxter house as the center of the Rebel center, the action was general all along the battle lines. Captain Oliver Basham with his two companies of Hill's Battalion stayed on his horse even after the battle began. An account in the Little Rock *True Democrat* stated that "Capt. Basham...was in command of his company and mounted, by permission, as he was too portly to walk. His men told him that he was too good a mark sitting on a horse and urged him to dismount. He took their advice, and just as his feet touched the ground and while his hand was yet on the pommel of the saddle, a bullet came, striking him on the knuckles, and which, had he been in the saddle, would have hit him plump in the abdomen."[40]

The Confederates successfully attacked and captured the Baxter house across the street from Federal Headquarters, taking it at the onset of the battle. As they entered the house they were confronted inside by a man known as Uncle Ben Davis, who was mentally deficient. He tried to stop the Rebels because of the danger to the women in the basement (Sarah Yeater and others), but the soldier who Uncle Ben grabbed, doubtless unaware of his condition, shot him. He died within the next day or two.

This Baxter house became the anchor of the Confederate center. The absent Pastor Baxter later described what he had learned of the action at his home:

> *[A] battle raged round my dwelling, and in sight of it men lay dead and dying. My house, being in full range of the enemy's fire, suffered from their cannonade, and I doubtless left none too soon for the safety of my family. The house which I formerly occupied near the College was seized and held for a time by the enemy, and was struck and pierced by more that fifty shot, shell and bullets....*
> *It was, and ever will be, a sad thought to me*

that this battle raged over the graves of my two children, and that the little mound, which covered the dust so precious tome, was doubtless trampled and defaced by that fierce soldiery. So dear, so hallowed was that spot to me that I cannot think of its sanctity being invaded by the fierce struggle which took place above where they sleep, without a sadness which words can not express.[41]

One Confederate soldier by the name of James Davis was wounded in the fighting. He was 43 years old and a resident of Fayetteville. He managed to get to his own home, where he collapsed, and died the following day. He is now buried in the Confederate cemetery not far from where the battle took place.

During the battle the Federal surgeons, Drs. Amos H. Coffee and Jonathan E. Tefft, "were very prompt in sending out their ambulance and directing where they should be driven, doing this while the engagement was still in progress."

There were various actions in different places. A Federal supply train of 10-15 wagons was overrun and destroyed by the Confederates, and a few escort soldiers were captured. The Rebels at one point took possession of the Federal hospital, capturing the men there, including Private James J. Hutchinson (or Hutchison) of the First Union Infantry. Harrison ordered Major Fitch on the right to send a company "to drive in the enemy's pickets at the hospital."

With battle raging with uncertain results, Harrison sent word to Lt. Col. Searle of the First Infantry "to hold his men in readiness to reinforce us at a moment's warning for I expected to need them within a half hour as I supposed the enemy to be changing position."

Before long, the Federals were advancing in an attempt to re-take the gains of the Rebels. Colonel Harrison reported:

> *At 8 a.m. our center had advanced and occupied the house, yard, and outbuildings, and hedges at my headquarters; the right wing had advanced to the arsenal, and the left occupied the open field on the northeast of town, while the enemy had possession of*

the whole hill-side east, the Davis [Baxter] place, opposite to, and the grove of, headquarters. This grove was formerly occupied by the building of the Arkansas College.[42]

Harrison decided to try to knock out the Confederate artillery, thereby depriving the Rebels of their best advantage. He sent his brother, Lt. Harrison, with an order for Lt Col. Bishop to get it done. Bishop then selected Lt. Robb of Company L to advance with two companies and try to silence the guns by picking off the individual artillerists with rifle fire.

Things were stalling for the Confederates. Cabell's opportunity to win this battle was fast dissipating. About 9:00 o'clock he ordered Colonel James C. Monroe's cavalry regiment, probably with Palmer's guerrillas attached to it, to mount for a charge on the Union center.

Union Dr. Carpenter was on the porch of a house on the Cassville Road behind Major Fitch's Federal right wing. "East of the road was a wide wooded ravine, in which, and screened by the timber, the enemy's cavalry formed for the charge," Carpenter recorded. "Suddenly I heard a tremednous yell, then the clatter of the hourses, then the toss of their flags, and then there upon us." The Confederates attacked up the Old Missouri Road (modern Dickson Street).

Dr. Seymour D. Carpenter. Author's Collection.

Carpenter went on to describe the event:

Major Ezra Fitch, who was in immediate command of the Battalion in front of me, I had always regarded as a dull and stupid sort of man. I particularly disliked him, because he always wore a tall black plume on his slouch hat, but he was somethig like a tortoise. He required coals to be put on his back

-58-

before he could get up a move. In the present instance he rose grandly to the occasion. As soon as he heard the yell, he rushed up and down in front of his line, brandishing a revolver in one hand, and the objectionable plume hat in the other, with oaths that would have done credit to the "army in Flanders," he admonished his men to stand steady, to reserve their fire until the enemy reached the brow of the hill, and then to "give 'em hell."

Colonel Harrison personally commanded the Federal center where the Confederate attack was aimed, and was in the neighborhood of the Union Cavalry's Company F when Monroe's cavalry attack came. He ordered the company to "fire low, take good aim and be sure to kill a man every time."

Dr. Carpenter continued his description:

The brow of the hill was about forty yards from the line. In a minute the long line of Cavalry appeared, the Major rushed in front, gave the command to fire, and a sheet of flame from five hundred carbines greeted them; dozens of men and horses went down; I could see the line waver, and the men frantically reining their horses, and swerving to the right and left. They were armed with sabres, and if they had pistols they did not use them. All our men had carbines, and revolvers, and in a minute not a Rebel was in sight, save the killed and wounded.... The Major sped the fleeting guests, with fresh vollets of oaths, and then he, and his men began giving assistance to the wounded. Not a man on our side had received a scratch. The whole affair was over in five minutes. It was a most thrilling sight, and for a moment I thought our men would be ridden down, which might have happened they had not charged in a single line.[43]

As the attackers went toward Major Fitch's Union troops ahead, they passed Federal troops at Federal Headquarters on their right. This put the Confederates into a "galling" crossfire from the front and right, "piling rebel men and horses in heaps." Quartermaster Sergeant S.D.

Haley of the Cavalry's Company D "especially distinguished himself in repelling the cavalry charge by Monroe." It was believed that Sgt. Haley fired the shot that brought down the Confederate color bearer, resulting in the capture of the flag.

On the right end of the Federal center, companies of the First Union Infantry were active in pouring in a crossfire on the Confederate charge. Captains William C. Parker of Company H, age 24, and Randall Smith of Company A, 29, were both in the forefront of the action "bravely cheering their men" when they each were wounded in the head. Neither wound was serious. Private James Shockley, one of Smith's men, "acted bravely" and fell mortally wounded. Private Shadrick Cockerill of the same was also killed. Privates George Bledsaw, Niles Slater and John Woods were all wounded.[44]

Colonel Harrison was in the neighborhood of the Union Cavalry's Company F when Monroe's cavalry attack came. He considered the attack to be "gallant and desperate," and ordered the company to "fire low, take good aim and be sure to kill a man every time." Fourth Sergeant William H. Baldy of Company L of Monroe's Regiment wrote afterward that "my horse was killed in the action at Fayetteville...whilst I was in execution of orders...."

The Colonel's younger brother Lt. Harrison at the time of the charge was on his way to Lt. Robb to deliver orders for him to bring up his reserves. He described the cavalry onslaught:

> *I saw coming over the rise in front of the Tebbetts house a charge of cavalry so splendid in its bearing, so daring it its on-rush, that the memory of it is as fresh in my mind as if it were but yesterday. Hastily seeking partial protection behind the tree, I looked with wonder, as well as admiration, upon that splendid body of horsemen as they swept down Dickson Street, and while the missiles of death were signing about me, there came to me the words of inspired Tennyson as he told of the Charge of the Light Brigade: "Ithe jaws of death, Into the mount of hell, Rode the six hundred."As this splendid body of cavalry came thundering down Dickson Street and when nearing College Avenue, they were met by a fire from the*

Federal soldiers behind the foundation wall, so deadly that the most heroic could not face it, and turning the corner onto College Avenue, the column moved on and disappeared by way of the lot [to the south of the Baxter House]....

[W]hile I stood by the tree as the cavalrymen came thundering down the road, many falling from their mounts, one horse (evidently wounded to its death) turned and with a terrific leap cleared the high plank fence and fell dead in the Baxter lot, carrying his rider with him, whom, though evidently wounded, freed himself from the dead horse and made his way around the house. As soon as the retreating squadron had disappeared from view, I, with several of my comrades, hastened to the corner where the dead and wounded were lying, and it was my good fortune to help them to a more comfortable position, bringing them water from a nearby well.[45]

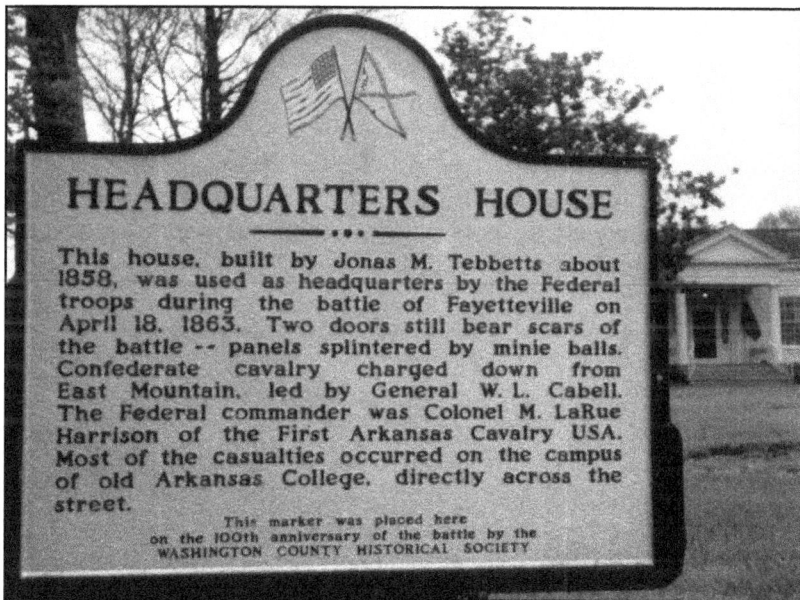

Historical Marker at Headquarters House.
Author's Collection.

Meanwhile, on orders from Lt. Col. Bishop, Lt. Robb was advancing with two companies from the Federal left to move in close upon the Confederate artillery "for the purpose of silencing if possible the enemy's battery." Robb formed his men in the field not far from the Davidson house. "Their artillerists and guns were out of sight, hidden by the brush," Colonel Harrison later explained, "and I ordered that after the discharge of their artillery my men should aim and fire their rifles about one foot above the blaze of the discharge."

Private Hugh Cook of Company L took this mission to heart, advancing 200 yards beyond his comrades. He shot two Confederates, captured their muskets and horses, and brought them back to Federal lines in the midst of the fighting. Lt. Col. Bishop made note of his actions in his after-battle report.

The fire of Robb's Federals was effective, and the Confederate battery lost one killed and several wounded, plus two horses killed and two wounded. Brigade Quartermaster Major Hugh M. Wilson was also badly wounded. By contrast, Company L of the First Union Cavalry lost only one casualty, when Sergeant Benjamin K. Graham was slightly wounded. "Two [Harrison said several] well directed volleys accomplished this object," Bishop reported of the attempt to silence the artillery. "At all events, the guns were limbered with great speed and hastily withdrawn to play no further part in the events of the day. I regard Lt. Robb's conduct as exceedingly daring & well timed." The artillery fell silent about 9 a.m., not long after Monroe's failed charge.[46]

General Cabell had a different version of the withdrawal of the artillery. He had wanted to move the artillery closer for more effective use but could not do so because he did not have adequate small arms to protect the guns. Later the artillery was withdrawn, he said in his report, because the supply of ammunition was exhausted. Credence is given this statement by his report a week later that his artillery guns still lacked ammunition. The truth of the matter is probably that the Confederates were indeed running out of ammunition just at the point that Federal rifle fire was getting very hot, and there was no longer any point in remaining any longer.

Whatever the reason, the only Rebel advantage in the battle was gone. In fact, all the Confederates momentum was now gone, but they persistently hung on to the Baxter house at the center of their line.

Both wings had been partially pushed back from this center point. Fighting continued for nearly an hour, with skirmishing, reconnoitering parties and stragglers.

Colonel Harrison said that during the battle "I walked about from place to place smoking cigars, talking and laughing with officers and men, encouraging them and singing patriotic songs." His 12-year-old son, Edward, encouraged him to stay behind cover because he exposed himself more than others and the regiment could not afford to lose him.

At about 9:30 a.m., he saw some of his men over at Headquarters House and went in after them. "I...found a number of stragglers from my own regiment skulking around, in and near this house, whom I arrested and turned over to Lieut. C.H. Wills with orders to take them and report to Major E. Fitch on the right," he reported. While there he also saw Captain John McCoy with his Company F of the First Arkansas Union Infantry and ordered him forward to support the middle in case of a renewed attack by the Confederates.

With the Confederate artillery the battle was left in the unequal contention between Federal and Confederate small arms. Cabell complained:

> *I found it impossible, with the arms I had, after my artillery ammunition was exhausted, to dislodge them from the houses and rifle-pits with the kind of arms my command had without losing all my horses and a large number of my men, as it was impossible to get near enough of them to make our aim effective without a great sacrifice of life, much greater than would have been justifiable under the circumstances....*
> *Had I had 500 long-range guns [muskets], with good cartridges, I could have taken the place in an hour. As it was, I could not advance my battery, as I had nothing to cover them with, as the enemy's guns were equal in range to the artillery. The Arkadelphia rifles, with the cartridges for them, are no better than shot-guns.*[47]

Faced with the reality that his attack on Fayetteville had failed, Cabell thought about burning the town to the ground. "I could have burned a large part of the town,: he reported afterward, but every house was filled with women and children, a great number of whom were the families of officers and soldiers in our service, and I did not deem it advisable to distress them any further, as their sufferings now are very grievous under the Federal rule."[48]

Instead, Cabell ordered the Confederates to retreat. "After a hard fight of three hours," recorded the return of Company C of Carroll's Regiment, "and finding that the enemy greatly outnumbered us, we retreated."

"[L]et it suffice to say," Lt. Col. Bishop wrote, "that at 10 o'clock a.m. it [the Confederate army] was a broken, disordered aggregate of galloping humanity, fleeing...for the Arkansas river." This was an over-statement. Contrary to Bishop's comment about "broken" and "disordered," Cabell reported that, "I withdrew my command in good order." Inasmuch as there was no pursuit by the victors, the Rebels undoubtedly did retreat in good order.

Cabell hoped the Federals would pursue him into the woods so he could attack them outside their houses and rifle-pits, but they did not accommodate him in this regard. Harrison wisely saw no sense in sending soldiers out on foot to chase Confederates on horses. If there was ever a victorious commander who was justified in not pursuing a defeated enemy, it was Harrison. About the best he could do was send out Captain Hopkins with Company I of the Cavalry to reconnoitre and drive in Rebel pickets. By noon the Confederates had given it up altogether and were in full retreat for Ozark.

The Battle of Fayetteville, Arkansas, was over.

The Aftermath
of the Battle

"[T]he rebels retreated," recalled Sarah Yeater, "leaving their dead on the ground.... Nine men dressed in Confederate gray were picked up in the yard and given decent burial in the cemetery.... [O]n examining my room in the cottage we found that two bullets had passed through the mattress where [baby] Charley and I had been lying, the window and chamber set were broken and several bullets had passed through the walls. In the large house many windows were broken and two cannon balls besides numerous bullets had left their marks."

The Confederates took with them all of their wounded that they could, but those too seriously hurt had to be left behind. Private J. O. Lively of Company H, Carroll's Regiment, went with them for a short distance, but could not go further. He was of necessity left at the White River, and later died of his wounds. It is likely that other wounded men also did not make it all the way back to Ozark. Privates N.M. Whorton of Company E and Thomas Sykes of Company I, both of Carroll's Regiment, were sent home to recover.

A considerable number of men also were doubling up on horses or walking. "I regret to say that I lost a good many horses," Cabell lamented. "The enemy's sharpshooters killed a good many with their long-range guns, and a few men left in charge of the horses evidently deserted them." Cabell estimated that he lost 200 horses killed, taken and stampeded, fifty of which were reported as captured by the Union men. That was more than one-fifth of the total.

To care for his wounded, General Cabell left behind a medical team under the command of Dr. Algemanus S. (usually known as A.S.) Holderness, a 29-year-old physician from Calhoun County, Arkansas. Born in North Carolina, he had graduated in 1854 from the medical department of the University of Pennsylvania. Moving to Arkansas to open his practice, he responded to the Confederate call to arms in 1861 and was ultimately assigned as the assistant surgeon in Monroe's Cavalry Regiment.

Confederate Casualties

No list of Confederate casualties by name has ever before been published. Most of these names were obtained by reviewing the individual soldier files of the regimental records.

Carroll's First Arkansas Confederate Cavalry

Company	Name	Rank	Wound
B	Clayton, R. P.	Private	Wounded
B	Hamilton, William	Private	Killed
B	Hargis, M. F.	Private	Killed
B	Hill, Claiborne	Private	Mortally wounded
C	Todd, William	Private	Wounded, captured
D	Price, D. C.	2nd Lt.	Wounded forearm
D	Ragan, E. L.	2nd Lt.	Wounded
D	Shropshire, Henry C.	Corp.	Wounded captured
E	Reed, George A.	3d Lt.	Wounded sent home
E	Whorton, N. M.	Private	Wounded sent home
F	Brandon, ---	Private	Killed
G	Jefferson, T. P.	Captain	Wounded, captured
G	Woods, A. P.	1st Sgt.	Killed
H	Hubbard, James	Private	Killed
H	Kindrick, O.	Private	Missing
H	Lively, J. O.	Private	Mortally wounded
H	Mathews, A. J.	Private	Wounded, captured
I	Dollins, F. D.	Private	Wounded
I	Harris, Robert	Private	Wounded
I	Sykes, Thomas	Private	Wounded, sent home
Battery	Hughey, William M.	Lt.	Wounded in arm

Dorsey's Missouri Cavalry

Ashcom, Benjamin H.   Wounded three times

Texas Cavalry

One man killed, Company A, Parson's 12th Texas Cavalry
One man wounded, company and regiment unknown

Monroe's First Arkansas Confederate Cavalry

Company	Name	Rank	Wound
B	Ford, P.	Private	Wounded, captured
B	Jones, H. J.	Private	Killed
B	Levillian, John	Private	Wounded, captured
B	Reddan, J. K.	Private	Missing, deserted
B	Rowan, James	Private	Killed
B	Touchstone, William	Private	Mortally wounded
C	Aaron, Arcillus	Corporal	Wounded, captured
C	Jackson, John R	Private	Wounded, captured
C	Savage, L. J.	Private	Wounded, captured
D	Garrett, John	Private	Missing
D	Holcomb, J. D.	Private	Wounded, died
D	Pauley, Andrew	Sergeant	Mortally wounded
E	Childress, G. H.	Private	Killed
F	Hall, H.	Not stated	Missing, deserted
F	Kain, J.	Not stated	Missing, deserted
G	Butler, J. P.	Private	Wounded, captured
G	Lissenberry, George	Private	Wounded, captured
G	Loretz, A. W.	Private	Killed
G	Williams, R. J. C.	Private	Wounded, captured
K	Brazil, T. S.	Private	Wounded, died
L	Bates, Marion	Sergeant	Wounded, captured

Palmer's Partisan Rangers

Hewitt, Alexander	Private	Wounded, captured
Hughes, Thomas	Corporal	Wounded, captured
Reager, Milton	Private	Wounded, captured
Reed, George W.	Private	Wounded, captured
Reynolds, William	Private	Mortally wounded
Teevan, George	Private	Wounded, captured
West, John	Private	Mortally wounded

Hill's Arkansas Battalion

Basham, Oliver	Captain	Wounded in hand

Staff of General Cabell

Brigade QM	Wilson, Hugh G.	Major	Wounded in hand
Regiment unknown -	Davis, James		Mortally wounded

Union Casualties

These names are based upon the battle report of Colonel Harrison and upon a review of the individual soldier files of the regimental records of the National Archives.

First Arkansas Union Cavalry

Company	Name	Rank	Wound
A	Burrows, Reuben B.	Pvt.	Killed
A	Fears, Josiah	Corpl	Slightly wounded
A	Hayes, John	Private	Severe wound arm
A	Jack, James	Private	Severely wounded
A	Kise, Frederick	Sergeant	Wounded slightly
A	Nail, Jesse	Private	Deserted
A	Riddle, Jonas	Private	Killed
B	Hottanhour, Gustavus	Lieut	Captured, paroled
B	Rutherford, John	Private	Captured, paroled
B	Scaggs, Hannibal	Private	Captured, paroled
C	Reed, Robert	Private	Captured
C	Wooten, William	Farrier	Slightly wounded
D	Asbill, John	Private	Severe wound chest
D	Carter, Adam	Private	Captured, paroled
D	Lewis, Henry C.	Corporal	Slightly wounded
D	Miller, William	Private	Captured
D	Quinton, William J.	Private	Slightly wounded
D	Strickland, Levi	Private	Captured, paroled
D	Temple, Francis M.	Private	Slightly wounded
E	Burrow, William M.	1st Sgt	Mortally wounded
E	Grubb, John A.	Private	Slightly wounded
E	Harp, John A.	Private	Wounded
E	Taylor, Jordan	Private	Severely wounded
G	Davis, William F.	Private	Severe wounded head
G	Morris, George A.	Sgt	Slightly wounded foot

First Arkansas Union Cavalry (Continued)

Company	Name	Rank	Wound
G	Russell, George W.	Private	Captured, hanged
H	Amos, Lucien	Private	Killed
H	Blevins, Elias	Corpl	Wounded, disabled
H	Davis, George W.	Private	Mortally wounded
H	York, William J.	Private	Severe wound foot
I	Bell, James D.	Private	Killed
I	Gregg, Allen C.	Private	Deserted
I	Manes, Jesse	Private	Deserted
I	Oxford, Jacob	Private	Captured, paroled
I	Sisemore, George W.	Private	Deserted
I	Steward, Joshua	Private	Killed
L	Graham, Benjamin K.	Sgt	Slightly wounded
M	Chyle, Davis	Private	Wounded
M	Johnson, William S.	Captain	Wounded arm
M	Norris, Doctor B.	Corpl	Slight wound head
M	Todd, Elijah	Private	Deserted

First Arkansas Union Infantry

Company	Name	Rank	Wound
A	Bledsaw, George W.	Private	Wounded
A	Cockerill, Shadrick	Private	Killed
A	Hutchins(on), John J.	Private	Captured, paroled
A	Shockley, James	Private	Mortally wounded
A	Slater, Niles	Private	Slightly wounded
A	Smith, Randall	Captain	Slight wound head
A	Woods, John	Corpl	Slightly wounded
E	Bingham, Thomas	Corpl	Deserted, reduced
E	Rupe, Daniel	Private	Slightly wounded
F	Rockdey, William	Private	Severely wounded
H	Nolen, Nathaniel	Private	Slightly wounded
H	Parker, William C.	Captain	Slight wound head

The 2nd Kansas, 11th Kansas, 27th Wisconsin and 37th Illinois were present on wagon train escort duty. They took no part in the battle, but a few of their men were captured.

A team of men was detailed to work with him as nurses. From Monroe's Cavalry, these included Privates Levi Miles of Company A, W.L. Tomlin of Company B, L.J. Savage of Company C, Wayne O. Worthington of Company D, George W. Lissenberry of Company G, Jason Tyson of Company K, and Alfred L. Killian of Company K. Private John Tennison of Palmer's partisan ranger company was also left behind as a nurse.

The dead and wounded of both sides were collected by the Union troops. "A wounded Rebel was brought to the porch where I was," said Dr. Carpenter. "He was making loud complaint; his middle finger had been shot off. I hastily dressed the wounded, and in a minute or two he dropped over dead. I was quite astounded, but on making a closer examination, found that he had been shot in the abdomen, and had died from internal hemorrhage."

The following day a burial team under the command of Captain William A. Alexander of Company C of Monroe's Regiment was sent back under a flag of truce. He carried this note:

> *Col. M. La Rue Harrison,*
> *Commanding Post of Fayetteville:*
>
> *Sir: The bearer of this letter, Captain Alexander, visits your post under a flag of truce, to bury any of my command that may be left dead from the engagement yesterday. I respectfully request that you will suffer him to get up the dead and wounded, and that you will extend to him such assistance as may be necessary to enable him to carry out my instructions.*
> *I am, sir, very respectfully, your obedient servant,*
> *W. L. Cabell, Brigadier General,*
> *Commanding Northwest Arkansas*

Colonel Harrison responded with the following note, and generously returned the battle flag that had been captured in Monroe's cavalry charge:

> *Brig. Gen. W. L. Cabell, Commanding:*

General: In reply to dispatches from you by hand of Captain Alexander, bearing flag of truce, I would respectfully state that the dead of your command have all been decently buried in coffins. The wounded are in charge of Surgeons Russell [Dr. Ira Russell of Massachusetts, a Unionist doctor in Fayetteville during much of the war] and Holderness, having been removed to our general hospital by my order. They are receiving every attention that men can receive - abundance of medicines, surgical instruments, and subsistence stores having been placed under the control of your surgeons. Rest assured, general, that your wounded shall receive the best of care, such as we would hope to have from you were we placed in a like situation.   Under the circumstances I consider it unnecessary to retain your flag, and therefore return it. Your prisoners shall be paroled, and as fast as the men whose names are mentioned in your list report to our lines, the exchange will be made.
I am, general, very truly, yours,

M. La Rue Harrison,
Colonel, Commanding.

Colonel Harrison was justifiably proud of his command. As soon as he was able on the day of the battle, he sent out a preliminary report by telegraph to his superiors:

April 10, 1863
Major General Samuel R. Curtis,

Arkansas is triumphant! The rebels...attacked Fayetteville at daylight this morning, and, after four hours' desperate fighting, they were completely routed, and retreated in disorder toward Ozark.... Our stores are all safe; not a thing burned or taken from us.
Every officer and man in my command was a hero; no one flinched.
M. La Rue Harrison
Colonel, Commanding, Fayetteville[49]

"Every field and line officer," Harrison wrote the following day, "and nearly every enlisted man, fought bravely, and I would not wish to be considered as disparaging any one when I can mention only a few of the many heroic men who sustained so nobly the honor of our flag." He went on to mentioned several officers by name: "Lieutenant-Colonel Bishop and Majors Fitch and Hunt...led their men coolly up in the face of the enemy's fire, and drove them from their position.... Lieutenant Roseman, Post-Adjutant, and Lieutenant Frank Strong, Acting Adjutant, First cavalry, deserve much praise."

Colonel Harrison reported Federal losses as 4 killed, 26 wounded, 16 prisoners and 35 missing. A thorough study of the records gives somewhat different figures, in part due to wounded men dying and missing men becoming accounted for. More accurate figures are 10 men killed and mortally wounded, 28 wounded but not mortally, and 26 captured (a figure from Cabell) and paroled. Cabell did not take any prisoners with him, but rather paroled them all. This made a total of 64 casualties, or about 6% of the Federal force.[50]

A week after the battle, General Cabell estimated his losses as not more than 20 killed, 30 wounded and 20 missing.[51] With a week to gather information, and in view of the incomplete records available now, this is probably about as close a number as can be determined.

Among the captured Confederates, Harrison said his men had captured the wounded Major Hugh G. Wilson (the Confederate brigade commissary), the wounded Captain T.P. Jefferson of Company G of Carroll's Cavalry, four sergeants, three corporals, and 46 privates.

Sarah Yeater said that there were nine dead Confederates picked up in the yard of the Baxter house alone, and Dr. Carpenter gave the same number of nine as being killed in Monroe's charge. Harrison said not less than 20 killed and 50 wounded were left in Fayetteville, and that citizens reported to him that many wounded were moving with the command. Dr. Carpenter said 50 to 60 were wounded in Monroe's charge alone. Eighteen dead can be identified by name. Likewise, 26 specific men can be identified as wounded (not mortally) and five more as missing. No unwounded Confederates prisoners are known by name, but Harrison said the Federals captured 54 men, "a part of them wounded." Some, then, were captured and unwounded.

Union Victory Proclamation by Colonel Harrison

Comrades in Arms: Let April 18, 1863, be ever remembered. The battle of Fayetteville has been fought and won. To-day the brave and victorious sons of Arkansas stand proudly upon the soil which their blood and their bravery have rendered sacred to every true-hearted American, but doubly sacred to them. In the light of this holy Sabbath sun we are permitted, through God's mercy, to gather together in His name and in the name of our common country, to offer up our heartfelt thanks to the Giver of every good and perfect gift, for the triumphs of our arms and for the blessings which we this day enjoy.

When yesterday's sun rose upon us the hostile hordes of a bitter and unprincipled foe were pouring their deadly fire among our ranks; the booming of his artillery was re-echoing from mountain to mountain, and the clattering hoofs of his cavalry were trampling in our streets. At meridian, General Cabell, with his shattered and panic-stricken cohorts, was retreating precipitately through the passes of the Boston Mountains toward the Arkansas River, leaving his dead and wounded in our hands.

Fellow-soldiers, it is to your honor and credit I say it, he could not have left them in better hands. Not one act of barbarity, or even unkindness, stains the laurels you so proudly wear. Such may your conduct ever be brave and unflinching in battle, kind and generous to the vanquished. Abstain from all cruelty and excess; respect the immunities of private property; never insult or injure women or children, the aged, the sick, or a fallen foe. Let us show to our enemies that the Federal soldiers of Arkansas are as generous as they are brave and patriotic; let us prove to them the justice of our cause the purity of our purpose, so that soon we may gather together under the broad folds of our time-honored and victorious banner every true-hearted son of Arkansas.

Fellow-soldiers, I congratulate you all upon the glorious victory you have won, by your cool and determined bravery, for that Union which our revolutionary sires established by their valor and sealed with their blood. More than all I do congratulate you that this battle was fought upon Arkansas soil, and this victory won by Arkansians alone, thereby testifying to our patriotic brethren in arms from other States that we are not only willing but anxious to second their efforts in rescuing our State from the dominion of traitors; but in all our rejoicing let us not neglect to shed the tear of regret over the graves of those heroic men who fell beside us fighting bravely for the nation's honor....

How many of the 54 captured are included in Cabell's wounded and missing figures cannot be known. Probably few of those missing a week afterward turned up later. It is reasonable to say that the Rebels suffered 70 to 100 casualties.

The total casualties in the Battle of Fayetteville, then, ranged from 135 to 165. Over three hours, this was one casualty about every 75 seconds. This represented approximately 6% of the Federals and 11% of the Confederates.[52]

* * *

In the glow of victory, Colonel Harrison waxed poetic. A Victory Proclamation was issued on Sunday, the day after the battle, which was read at religious services.

Meanwhile, congratulations from Union commanders came pouring in to the justifiably self-satisfied Union soldiers of Arkansas. The initial word that went out from Fayetteville was that the place had been captured by Cabell. "The report which reached here two days ago, that Fayetteville had been taken on the 18th instant by a rebel force of fifteen hundred men, under command of General Cabell," wrote diarist Wiley Britton of the 6th Kansas Cavalry, "turns out to be untrue." Colonel Phillips reported of Harrison that, "My first report was he was taken in."

The fact that Arkansas' own Unionist sons had attained the victory over Arkansas Confederates was immediately recognized. Wiley Britton noted in his diary that, "The loyal Arkansas soldiers are represented to have acted with distinguished bravery throughout the contest."

General Herron, who had scathingly criticized the First Arkansas Union Cavalry four months earlier and had shot a retreating member of the regiment, was just as quick to send his compliments:

> *Headquarters Army of the Frontier,*
> *April 19, 1863*
>
> *Col. M. La Rue Harrison*
> *Commanding at Fayetteville*

> *I must congratulate you on the success of yesterday. It augers well for the future of Arkansas when her loyal troops have beaten the enemy in their first encounter. Such success should encourage us, and I hope soon to see 10,000 loyal men of Arkansas arrayed on the side of the Union. You have nobly sustained yourselves, and deserve a country's gratitude.*
> *F. J. Herron, Major-General[53]*

The commander of the entire department, General Curtis, added his respects:

> *Department of the Missouri,*
> *April 20, 1863*
> *Colonel Harrison, Fayetteville, Ark:*
> *Dispatch of yesterday received. Tender my thanks to the soldiers of your command for their gallant conduct in the battle of Fayetteville. You have done nobly. Arkansas vindicates her own honor by repulsing the rebel flag with her own brave sons. Send minute reports, naming the most deserving officers and men.*
> *Saml. R. Curtis, Major-General[54]*

One of the inevitable consequences of the battle, however, was dealing with those men whose performances were not up to expectations. Harrison's initial telegram notwithstanding, some of the men did flinch. Doubtless with the full knowledge and approval of Colonel Harrison, regimental commander Lt. Col. Bishop moved quickly to discipline some non-commission officers. Two days after the battle the following order was issued:

> *Headquarters, 1st Ark Cav Vols*
> *Fayetteville Ark, April 20, 1863*
> *Special Order No. 48*
>
> *Corporal John Reed of Squadron [Company] "A" is hereby reduced to the ranks for cowardice in the face of the enemy during the battle at this place on the 18th inst. With few exceptions, the regiment behaved*

> *nobly and the commanding officer takes this occasion*
> *to say that he could not have been better pleased with*
> *the daring valor of the men under a murderous fire.*
> *By Order of A W Bishop*
> *Lt Col 1st Ark Cav, Commanding*

The next day, April 21st, Sergeants Joseph Bell of Company I and Cyrus Barber of Company L were also "reduced to the ranks for cowardice in the face of the enemy on the 18th...." On the 22nd, Corporal Jerome Bays was "busted" to private. He deserted less than four months later. Conversely, Private Elias Blevins was promoted to Sergeant in Company A for his bravery in battle.

The First Arkansas Union Infantry had its own disciplinary work to do. Corporal Thomas Bingham of Company E was reduced to private. Crittenden C. Wells, the Lieutenant who "disgracefully ran away" when the battle began, earned specific mention in Colonel Harrison's report and was in immediate trouble with his superiors. Major Ham called him incompetent and filed court martial papers for cowardly conduct in the face of the enemy. Seeing the handwriting on the wall, Wells resigned his commission. On July 31, 1863, he was dishonorably dismissed from the service by order of Secretary of War Edwin M. Stanton.

***

Rather than receiving congratulations, General Cabell had a much less enviable task of explaining the failure to take Fayetteville to his superiors. "I did not take the place," he wrote to General Steele, "and if I had with me every man I have on paper, armed as they are [with shotguns], I could have done no more. I made an honest effort to take the place, and have given them a severe blow, and one that will prove to be a good one in the end, as it will curb their utter lawlessness...."[55]

Cabell felt ambivalent about the performance of his men in the battle. "The troops, with few exceptions, all fought well," he said, "and are now [seven days later] in fine spirits, ready and willing to try the enemy again." In another place he said "the officers and men, with a few exceptions, acted well." On the other hand, in another section of his report he said, "I had too many inefficient officers and not enough long-range guns." It's a tough balancing act to lose a battle, explain why to a superior officer and yet be fair with your men and keep up their morale.

Cabell took some solace in the battle, even though he was not successful:

> *Although I did not capture Fayetteville and drive the enemy out from it, yet my expedition will prove to be a beneficial one, a it will in future curb the lawlessness of the troops there; will cause them to send all their regular troops east, and it will keep the place in that condition in reference to numbers that will enable me, with a small increase of my force, and with a few hundred long-range guns, to take the place. Besides this, I have obtained information that cannot be obtained from any other source, as it is impossible to get correct information from people living there. Our friends are all too anxious to rid the country of their presence to state things as they really are. I find this to be true in every respect but in reference to artillery. Our enemies (Union men) will give no information at all, either in reference to the enemy or country.*

The Confederate forces in northwest Arkansas were hurting after the loss at Fayetteville. Just three days after the battle, on April 21st, General Cabell wrote to Lt. Gen. Holmes "that, being entirely out of forage for my horses and bread for my men, I will be compelled to join General Steele. My horses have had but one feed of corn since the morning of the 17th, and that one-half ration...." In a very frank letter, Cabell continued, "There is nothing north of the mountains to subsist either men or horses.... But, sir, no man can concentrate mounted troops and keep them together when there is no forage, no horseshoes, nothing to make them efficient."[56]

***

Not knowing the desperation of the Confederate position, the Federals let their imaginations run away to conclusions of imminent Rebel attack. Five days after the battle, Union commander Colonel Harrison wrote to General Curtis:

> General Curtis:
>
> Can we be re-enforced, and that immediately? We can never hold this place without artillery and horses. There is no use in disguising the fact. Last night I was positive that Cabell and the Fort Smith Indians had combined to attack me at daylight. My men stood under arms from midnight until after sunrise. Such an attack is brewing, and will come in force in a few days. We have no stores here; we have nothing to eat, and cannot get out trains, with good luck, until the 28th. Must we starve, and then have all the conscripts surrender to an overwhelming force, that will shoot them as deserters? We haul forage 45 miles, and weaken our command by large escorts. We can make no reconnaissances nor scouts for want of horses, and could not protect our rear and flanks in a retreat. The enemy are splendidly mounted. The men are brave, and have achievbed a splendid victory, but we must have help or fall back. Answer immediately what I sahll do. Colonel Phillips is about 90 miles from here, and of no use to us in case of an attack from Fort Smith. I should have to face the enemy's artillery all the way to get there.
>
> M. La Rue Harrison[57]

General Curtis decided to pull back the most exposed Federal positions. "[W]e have too much to bear from organized, armed, and equipped rebel forces now to allow our forces to range far from central positions, where succor can be certain, and effective resistance secured," Curtis wrote to General Blunt. "Phillips must come back, and, I think, Harrison, also, at least until they can be more fully equipped, and a reserve force is more completely massed in a central location."

And so the order went out for the Federal outpost at Fayetteville to be evacuated. Equipment, supplies and other items were loaded into wagons, and started out for Springfield, Missouri, on the afternoon of April 24th, just six days after the successful battle to defend the town. The evacuation was completed by the following afternoon. Lt. Col. Bishop described it:

> On the afternoon of the 25th or April the
> dreaded evacuation began.... At three o'clock the
> motley assemblage began to move; the First Arkansas
> Cavalry, dismounted, (for their horses had been worn
> out in the service,) and with transportation altogether
> insufficiency; the First Arkansas Infantry with no
> transportation at all; and in their rear, preceding a
> rear guard, a citizen train bearing and accompanied by
> nearly two thousand people.
> This last feature was especially distressing.
> Family after family moved despondingly out; the father
> careworn and dejected; the mother anxious, yet
> patient, and the children with a curious mixture of
> wonder and excitement that served to buoy up rather
> than depress. All were in the greatest destitution. The
> rude cart pulled wearily along by the half famished
> oxen, or the rough wagon with it's tattered covering,
> contained all the worldly effect that they had the means
> of rescuing from plundering rebels... But enough. A
> brighter day will dawn for Arkansas.

Confederate General Cabell reported of the enemy withdrawal saying that "A train of at least 100 wagons left Fayetteville a few days ago filled with Union families." Even Pastor Baxter left this time, not only because he did not want to stay under Confederate control, but also because of Federal "soldiers whose conduct rendered my position any thing but agreeable."

Union Dr. Seymour Carpenter was left in command of the hospital in town as the soldiers pulled out. For days no Confederates entered to take up the abandoned gift of Fayetteville that had been left to them. "At length, on the eleventh day," Carpenter recorded, "a company of ragamuffins, under the command of one Capt. Palmer, who styled himself a 'Partisan Ranger,' appeared."[58] There was justice and irony in this, for Palmer and his "ragamuffins" had been active participants in the Battle of Fayetteville on April 18th. General Cabell soon crossed the Boston Mountains and occupied the town, too. Colonel John Scott became the commander of troops headquartered in Fayetteville, using the very same Tebbetts home as his headquarters.

But it was not for long. Colonel Harrison and the Arkansas Federal soldiers returned and re-occupied Fayetteville on September 22, 1863, and, so long as the war continued, never again left again.

Conclusions about the Battle of Fayetteville

The Battle of Fayetteville, Arkansas, took place at the exact mid-point between the firing upon Fort Sumter in April of 1861 and the surrender of Robert E. Lee in April of 1865. As a struggle on Arkansas soil between Arkansas Confederates and Arkansas Federals it epitomizes just what the Civil War was all about. In a way it was, in a single morning, a microcosm of the entire war.

 The best summary of what the Battle of Fayetteville meant for the Arkansas Federals was stated by Lt. Col. Albert W. Bishop of the First Arkansas Union Cavalry. Although always partisan in his writings, he was indisputably accurate when he wrote:

> *This engagement, though of minor importance as compared with the contests of the Army of the Potomac, or the struggles that have recently culminated in the capitulation at Vicksburg, is not without its significance. It was the first battle of the war in which the loyal men of Arkansas were alone opposed to the organized treason of the State, and gave a very decided reproof to the rebel slander, that the Union men of Arkansas will not fight.*[59]

In a letter on May 9, 1863, to General Curtis many of the officers of the First Arkansas Union Cavalry reported that the victory at Fayetteville "gave renewed confidence to the loyal Arkansians and impetus to the recruiting service in the State of Arkansas."

Even more significant than the Unionists claiming that this lesson was taught was the concession of the Confederates that the lesson had been learned. General William L. Cabell gave a very great compliment to the Arkansas Federals after the Battle of Fayetteville:

> *The enemy all (both infantry and cavalry) fought well, equally as well as any Federal troops I have*

ever seen. Although it was thought by a great many that they would make but a feeble stand, the reverse, however, was the case, as they resisted every attack made on them, and, as fast driven out of one house, would occupy another and deliver their fire.[60]

The Confederates are also entitled to their claim of success. It should not be forgotten that Cabell's purpose in attacking Fayetteville was to make a mere "dash," to probe its defenses to see if it would crumble. He did this. The defenses stood firm and he withdrew. It was never intended to make a massive, desperate assault, but no one can say that 100 casualties in 180 minutes was a half-hearted effort.

Yet the Confederates did accomplish two things in their attack. First, as Cabell stated, "Although I did not capture Fayetteville and drive the enemy out from it, yet my expedition will prove to be a beneficial one, as it will in future curb the lawlessness of the troops there...." In his book after the war, Colonel John M. Harrel wrote about the battle and the victory proclamation of Colonel Harrison. "The address was framed upon a high and familiar precedent," and was altogether in a tone honorable to the piety and patriotism of its author," he wrote generously. "It may have been the restraining influence of these sacred feelings, and not the march of Cabell, which caused the cessation of the pillage and murder that had been indulged in by the triumphant defenders of Fayetteville, or by their agents. The fact is, thenceforth they were discontinued, and comparative quietude resumed its sway among these romantic valleys."

Second, General Cabell did prove to be the temporary victor in the dispute over Fayetteville. Although he lost the actual battle, he won the campaign. Within a week of the attack, his Confederates occupied the town just as surely as if they had taken it in the fight on April 18th. Because of Cabell's attack, the Union generals felt took weak and exposed to keep a force there and therefore ordered the abandonment of Fayetteville. Yet, in turn, it must also be said that if Colonel Harrison won a battle but lost a campaign, he also won the war. The Federals returned in September of 1863, never to be driven from Fayetteville again so long as the Civil War continued.

Both sides, then, can fairly claim their respective victories in the Battle of Fayetteville.

The Fate of the Participants in the Battle of Fayetteville

Brigadier General William L. Cabell, who became known as "Old Tige," fought with his brigade until the war ended for him on October 25, 1864, when he was captured by Union forces while engaged in Sterling Price's great raid into Missouri. He was held in a prisoner of war camp until the surrender of Confederate forces the following year. In July of 1865, he signed an oath of allegiance to the United States and became a free man again.

Cabell practiced law in Fort Smith, Arkansas, after the war, and in 1872 moved on to Dallas. There he became a successful businessman and politician, serving several years as mayor. He was active in the United Confederate Veterans until his death on February 27, 1911. He is buried in Greenwood Cemetery in Dallas, Texas.

Colonel James C. Monroe was universally admired by the Union defenders at the Battle of Fayetteville. Union Lt. Col. Bishop correctly said that "Colonel Monroe, a brave and gallant officer, was an especial reliance" of General Cabell. Monroe fought on through the war, and was with General Cabell at the time that the general was captured. He would have become brigade commander himself at that time except that he was wounded at the time Cabell was captured. When the Confederate armies began surrendering across the South, Monroe could not reconcile himself to it. Rather than take the oath of allegiance, he went south to Mexico.

Many of the participants in the Battle of Fayetteville did not survive the war. Lt. Col. Sebron M. Noble was killed in the Red River campiagn in Louisiana in April 1864. One of his two company commanders, Captain Joseph P. Weir of Company A of the 18th Texas Cavalry, was killed a month later in action at Yellow Bayou, Louisiana.

Caleb Dorsey survived the war but left Missouri and became a rancher in California. He took his widowed mother and adult brothers and sisters and went to Stanislaus County southeast of San Francisco.

Artillery Captain William M. Hughey married in 1864. Lt. Col. Lee L. Thomson was promoted to full colonel to command Carroll's Regiment. Lt. Jim Ferguson, who captured the dancers on the edge of town, broke his leg and resigned from the Confederate Army in October of 1863 due to the resulting unfitness for duty. Dr. A .S. Holderness, who was designated to stay behind with the Rebel wounded, fought until the surrender in 1865, then settled in Dallas County, Arkansas, and became a very successful physician and large landowner. In 1882 he sold the land which became the location for the town of Fordyce.

On the Union side, soon after the Battle of Fayetteville Colonel M. LaRue Harrison applied to become a Brigadier General. "I claim to be an Arkansian because I espoused her cause in the darkest hour and have labored and fought for her for a year," he said in a letter dated May 14, 1863. "Moreover I hope and expect to make that state my home at the close of the war." Both of these were accurate statements, but he did not get the promotion for two more years. He continued on in command of the First Arkansas Union Cavalry in Fayetteville until the end of the war.

Captain Randall Smith of Company A of the First Arkansas Union Infantry, who was slightly wounded in the head in the battle, later formally accused Colonel Harrison of cowardice and dereliction of duty. He cited instances of claimed confusion and incompetence at the both the Battle of Prairie Grove and the Battle of Fayetteville. He "became so much frightened as to be unable and did not give orders to his men until the action had ceased." In his defense, Harrison set forth a detailed accounting of his behavior on both of those occasions. Officers from both the Cavalry and the Infantry regiments were offered as witnesses in defense of the charges. Although Harrison demanded a court martial to clear his name, this superiors were never concerned enough about the charges to formally look into them.

The matter came to an end when Captain Smith self-destructed. In Missouri in July of 1863 Smith was accused of "conduct unbecoming an officer and gentleman. He visited a private house in the city last night and conducted himself in a very disgraceful manner...." Smith was drunk and disorderly in the presence of enlisted men while on duty as the brigade officer of the day. He was dishonorably dismissed from the service of the United States. There was no further talk to court martialing Harrison. In fact, at the end of the war he was given the brevet rank of Brigadier General.

As he said he would, Colonel Harrison did indeed make Arkansas his home after the war. Many Northern Federal officers did this, and for a time they prospered in the Reconstruction (i.e., carpetbagger) era of the defeated South. He did the engineering work to lay out a community called Stiffler Springs in Boone County, which honored him by naming their town Harrison, Arkansas. He also became mayor of Fayetteville, an experience which unfortunately turned very sour. His actions became so unpopular with the residents that they successfully petitioned the State legislature to disincorporate the city to put him out of office. This was done, and Harrison's term as mayor was ended.

As Reconstruction began to run out, so did many of the Northerners. Harrison went back to Washington, D.C., to work in the postal service, and died there in 1890. He is buried in the Arlington National Cemetery just outside of Washington, D.C.

From the First Arkansas Union Infantry, Colonel James M. Johnson was also breveted as a Brigadier General. In 1864 he was elected to Congress from Arkansas, but he and all Arkansas representatives were rejected by the Republican Congress. Major Elijah D. Ham of the same regiment resigned from the Army in March of 1864 due to disability and the fact that he was elected to the newly constituted Unionist Arkansas Legislature as a state senator from Madison and Benton counties.

Lieutenant Colonel Albert W. Bishop served as Adjutant General of Arkansas at the end of the war. He also fell in love with Arkansas during his military duties, and when his military service ended he stayed on. In 1867 he became a bankruptcy administrator , and then treasurer and a trustee of Arkansas Industrial University. From 1873 to 1875 he was president of the University, which later became at the University of Arkansas. In 1876 he was the reformed Republican candidate for governor, but went down to defeat as the ex-Confederates regained control of the State. With the Reconstruction era over, he too went back north to Buffalo, New York, to practice law. He died in 1901.

Major Thomas J. Hunt of the First Arkansas Union Cavalry was no carpetbagger. He was a homegrown Southerner and he stayed in Fayetteville all his life, becoming a successful farmer and business leader after the war. He served two terms in the State Senate, and was postmaster of Fayetteville for eight years from 1889 to 1893 and from 1898 to 1902. Major Elijah D. Ham of the First Arkansas Union Infantry was also elected to the legislature while the war was still going, and resigned to take his seat there.

Major Ezra Fitch, who commanded the Federal right against Monroe's cavalry charge, resigned from the Army in November of 1864 due to physical disability resulting from a hernia. Prior to the battle he had been an object of dislike to Dr. Seymour D. Carpenter, but not afterward. "Even after the battle I had the most profound respect for the Major," the doctor wrote, "for it is not always safe to judge a man only by his looks." Dr. Carpenter went back to Cedar Rapids, Iowa, and later went on to Chicago. He was active in the railroad business, finding both great success and failure. He wrote his autobiography in 1907. Captain Rowan E. M. Mack of Company G, who protected the Federal right from being flanked, was killed in action in Washington County on May 28, 1864.

Grave of Captain Joseph S. Robb Fayetteville National Cemetery (Author's Collection.)

Lieutenant Robb, whose men fired on the Confederate artillerymen, did not live to see another year. He was promoted to Captain of Company L, but was accidentally shot in the leg and bled to death. An obviously distressed Colonel Harrison to the department ajdutant about it on Christmas Eve 1863. "I would most repsectfully report to you that Capt. Joseph S. Robb Co. L 1st Ark Cav died in hospital at this place at two o'clock a.m. 20th instant of accidental gunshot wound, ball passing between bones of the leg below the left knee cutting the main artery, causing excessive hemorrhage,"

Harrison wrote. "Very little is known of his relations except that his mother is a Mrs. Cynthias House of Palmyra, Iowa. Capt. Robb was a very gallant officer and tendered much important service to this government.... I believe him to have been thoroughly honest."

Lt. Elizur B. Harrison, the younger brother of the Colonel, also found a home in Fayetteville. He stayed on there for the rest of his life, working in banking and business and becoming a fixture in the community. He married Sallie Yeater, one of the women in the Baxter house who was spared when the artillery shell landed in the kettle of lye. Over the years he became the living memorial in Fayetteville of the battle that was fought there one day in April of 1863. He was present 63 years later at the 1926 dedication of a tablet at the corner of Dickson and College streets where Monroe's First Arkansas Confederate Cavalry clashed with the First Arkansas Union Cavalry. He died in 1929.

They are all long ago gone from us. All we can do now is look at the few photographs that remain of them, and read what they wrote. We can go where they were, and try to feel what they felt. On the quiet grounds of Headquarters House where the Battle of Fayetteville once raged, now, separated only by time, we can stand among them.

THE END.

Endnotes

1. Goodspeed, ed. *History of Benton , Washington, Carroll, Madison, Crawford, Franklin and Sebastian Counties, Arkansas,* (Chicago: 1889), p. 227-232.

2. Nash, *Biographical Sketches of Gen. Pat B. Cleburne and Gen. T. C. Hindman*, 190 ff\; John F. Maguire, "The Irish in America" (New York, 1867), p. 367.

3. Baxter, William, *Pea Ridge and Prairie Grove, or Scenes and Incidents of the War in Arkansas* (Cincinnati: Hitchcock & Walden, 1869), pp. 53-54.

4. Scott, Kim Allen, "Witness for the Prosecution: The Civil War Letters of Lieutenant George Taylor." *Arkansas Historical Quarterly,* Autumn 1989, pp 260-271.

5.U.S. War Department, *The War of the Rebelluion: A compilation of the Official Records of the Union and Confederate Armies,* 128 volumes (hereafter designated as *Official Records),* Series 1, Volume 8, pp. 70-71.

6. Baxter, pp. 175-176.

7. Baxter p. 192.

8. Walker, Edwin S., ed., *Genealogical Notes of the Carpenter Family* (Springfield, IL: Illinois State Journal Co., 1907), p. 140.

9. Baxter, pp 224-225.

10. Regimental Records, Confederate Regiments from Arkansas, First Regiment of Cavalry, under the Company B Returns, National Archives Microfilm.

11. Regimental Records, Confederate Regiments from Arkansas, 8th Field Battery, under the name of William M. Hughey, National Archives Microfilm.

12. *Official Records*, Series 1, Volume 22, Part 2, p. 604.

13. Bishop, Albert W., *Loyalty on the Frontier, or Sketches of Union Men of the Southwest* (St. Louis, 1863), p. 97.

14. Regimental Records, Confederate Regiments from Texas, under the name of each unit. National Archives Microfilm.

15. Woods, James M., *Rebellion and Realignment: Arkansas's Road to Secession* (Fayetteville, University of Arkansas Press, 1987), p. 160.

16. Bishop, pp. 88-104.

17. Bishop, pp. 11-12.

18. Bishop, Albert W., An Oration Delivered at Fayetteville, Arkansas, July 4, 1865 (New York: 1865).

19. *Official Records*, Series 3, Volume 2, p. 958.

20. Bishop, pp. 62-63.

21. *Official Records*, Series 1, Volume 22, Part 1, p. 72.

22. Regimental Records, Union Regiments from Arkansas, First Regiment of Cavalry, under the name of Marcus LaRue Harrison, National Archives Microfilm.

23. *Official Records*, Series I, Volume 22, Part 1, p. 137.

24. Tillie, Nannie M., ed, *Federal on the Frontier: The Diary of Benajmin F. McIntyre* (Austin, 1963), p. 58.

25. *Official Records*, Series 1, Volume 22, Part 1, pp. 102-103.

26. Regimental Records, Union Regiments from Arkansas, First Regiment of Cavalry, under the Regimental Returns, the Company M Returns, and the Company F Returns, National Archives Microfilm.

27. *Official Records*, Series 1, Volume 22, Part 2, p. 801.

28. *Official Records*, Series 1, Volume 22, Part 1, p. 312.

29. *Official Records*, Series 1, Volume 22, Part 2, p. 848.

30. *Official Records*, Series 1, Volume 22, Part 2, p. 192.

31. *Official Records*, Series 1, Volume 22, Part 2, p. 191.

32. Returns of U.S. Military Posts, 1800-1916, Fayetteville, Arkansas, April1863.

33. Regimental Records, Union Regiments from Arkansas, First Regiment of Infantry, under the name of Mathew W. Sumner, National Archives Microfilm.

34. *Official Records*, Series 1, Volume 22, Part 1, p.306. Regimental Records, Union Regiments from Arkansas, First Regiment of Cavalry, under the name of Marcus LaRue Harrison, National Archives Microfilm.

35. Bishop, Albert W., battle report, April 22, 1863, Regimental Records for the First Arkansas Union Cavalry, National Archives in Washington, D.C.

36. *Official Records*, Series 1, Volume 22, Part 1, pp. 306, 310.

37. Regimental Records, Union Regiments from Arkansas, First Regiment of Infantry under the names of Francis W. Cannon and Gilbert C. Luper, and the First Regiment of Cavalry, under the names of Thomas Bingham,. Cyrus Barber and Marcus LaRue Harrison, National Archives Microfilm.

38. Regimental Records, Union Regiments from Arkansas, First Regiment of Cavalry, under the names of William Johnson, Davis Chyle and Doctor B. Norris, National Archives Microfilm.

39. *Official Records*, Series 1, Volume 22, Part 1, p. 312.

40. Little Rock *True Democrat,* April 29, 1863, p. 1.

41. Baxter, pp. 225-226.

42. *Official Records*, Series 1, Volume 22, Part 1, p. 306-308.

43. Walker, pp. 142-143.

44. Regimental Records, Union Regiments from Arkansas, First Regiment of Infantry, under the names of William C. Parker, Randall Smith, James Shockley, Shadrick Cockerill, George Bledsaw, Niles Slater, and John Woods, National Archives Microfilm.

45. Harrison, E. B., "The Battle of Fayetteville," *Flashback* (Washington County Historical Society) (May 1968), Vol. 18, No 2, pp. 18-19.

46. Bishop, Albert W., battle report, April 22, 1863, Regimental Records for the First Arkansas Union Cavalry, National Archives in Washington, D.C.

47. *Official Records*, Series 1, Volume 22, Part 1, p. 311.

48. *Official Records*, Series 1, Volume 22, Part 1, p. 311.

49. *Official Records*, Series 1, Volume 22, Part 1, p. 305-306.

50. *Official Records*, Series 1, Volume 22, Part 1, pp. 306-308.

51. *Official Records*, Series 1, Volume 22, Part 1, pp. 306-308.

52. *Official Records*, Series 1, Volume 22, Part 1, pp. 306-308.

53. *Official Records*, Series 1, Volume 22, Part 1, p. 309.

54. *Official Records*, Series 1, Volume 22, Part 1, p. 310.

55. *Official Records*, Series 1, Volume 22, Part 2, p. 829.

56. *Official Records*, Series 1, Volume 22, Part 2, p. 828-829.

57. *Official Records*, Series 1, Volume 22, Part 2, p. 246.

58. Walker, p.144.

59. Bishop, p. 217.

60. *Official Records*, Series 1, Volume 22, Part 1, p. 311.

Selected Bibliography

Allen, Desmond Walls, *First Arkansas Union Cavalry* (Conway, Arkansas, 1987).

Anderson, John Q., Editor, *Campaigning with Parson's Texas Cavalry Brigade, CSA: The War Journals and Letters of the Four Orr Brothers, 12th Texas Cavalry Regiment* (Waco, 1967).

Baxter, William, *Pea Ridge and Prairie Grove, or Scenes and Incidents of the War in Arkansas* (Cincinnati, 1864).

Bishop, Albert, *Loyalty on the Frontier, or Sketches of Union Men of the South-West* (St. Louis, 1863).

Bishop, Albert, *Report of the Adjutant General*, 1867.

Bishop, Albert, *An Oration Delivered at Fayetteville, Arkansas, July 4, 1865* (New York, 1865).

Campbell, William S., *One Hundred Years of Fayetteville, 1828-1928.* (Fayetteville, 1977).

Harrell, John M., *Confederate Military History, Volume 14, Arkansas* (Wilmington, 1988).

Harrison, Elizur B., "The Battle of Fayetteville," *Flashback*, Washington County Historical Society, Volume 18, No. 2, (May 1968).

Little Rock True Democrat, April 29, 1863

National Archives Records for (1) The First Arkansas Union Cavalry, (2) The First Arkansas Union Infantry. (3) Monroe's First Arkansas Confederate Cavalry, (4) Carroll's First Arkansas Confederate Cavalry (listed under Gordon's Cavalry, (5) Miscellaneous Arkansas Records, (6) Dorsey's Missouri Cavalry, (7) The 12th Texas Cavalry, (8) The 18th Texas Cavalry, and (9) Post reports. On microfilm and at the Washington D.C. office.

U.S. War Department, *The War of the Rebelluion: A compilation of the Official Records of the Union and Confederate Armies*, 128 volumes.

Walker, Edwin S., *Genealogical Notes of the Carpenter Family*, (Springfield, Illinois, 1907).

Yeater, Sarah J., "My Experiences During the Wat Between the States," *The Arkansas HistoricalQuarterly*, Vol IV, No. 1, (Spring 1945).

Washington Telegraph, April 29, 1863.

Index